Presentation Skills for Scientists and Engineers

Jean-Philippe Dionne

Presentation Skills for Scientists and Engineers

The Slide Master

Jean-Philippe Dionne
Gatineau, Québec, Canada

ISBN 978-3-030-66071-0 ISBN 978-3-030-66069-7 (eBook)
https://doi.org/10.1007/978-3-030-66069-7

This Springer imprint is published by the registered company Springer Nature Switzerland AG
The registered company address is: Gewerbestrasse 11, 6330 Cham, Switzerland

Contents

About the Author

Jean-Philippe Dionne, Ph.D. in Mechanical Engineering (McGill University, Canada), has authored a number of scientific publications (conference proceedings, journal or magazine articles) in the field of personal protective equipment during his 20-year career in the industry. He has spent countless hours preparing and delivering presentations for scientific conferences and other events, after he finally abandoned the good old projector and transparencies in the late 1990s.

Introduction

1

J.-P. Dionne, *Presentation Skills for Scientists and Engineers*,
https://doi.org/10.1007/978-3-030-66069-7_1

1.1 Introduction to This Book

A good speech starts with a good slide deck. This applies just as much to gifted speakers as to those wishing they could hide six feet under ground while on stage.

Great speakers are entertaining and instill confidence. But despite their speaking skills, their message might not get across properly if they rely on poor visual backup.

On the brighter side, even a poor public speaker with a shaky mastery of the language can deliver a great and well-structured talk, if backed up by the appropriate visuals.

It amazes me every time I attend scientific conferences, to realize that the vast majority of these high-level and experienced scientists, definitely stars in their respective domains of research, have no clue when it comes to delivering a presentation. Over and over the same basic and obvious mistakes, which make the audience either stare at their watches hoping for a quick end to this torture or fight against the humiliation of snoring in public.

There is nothing wrong with spending 10 years of one's life at the university to get entitled to these few additional letters at the end of one's name. But why isn't any time spent on learning how to give an effective presentation?

Whether you are a scientist, an economist, an engineer, a medical doctor, a politician, everything starts with a good slide deck.

I am confident that both the young undergraduates with very little presentation experience and the older more seasoned scientists with possibly less up to date "computer skills" are likely to benefit from this book.

The following pages are filled with numerous tricks and advice written in a very concise manner. Simple, right to the point.

(You are expected to already know how to use a slide presentation software – only a few instructions on how to use specific features of PowerPoint are included)

1.2 Get Inspired by Documentaries

An on-screen slide presentation is to a documentary what a play on Broadway is to a Hollywood movie. A slide presentation is more up close and personal. The spectators are closer to the action and even part of it, as the actors do interact with the audience. On the other hand, a movie is less personal, but filled with exciting special effects.

Just like there is still room and interest nowadays for plays, despite their primitive technologies compared to movies, there are still some unique and desirable features in a presentation compared to a documentary.

Yet I urge presenters to get inspired by the content of documentaries, as those are typically created by highly skilled professional teams having access to the greatest tools. But even more important, get inspired by what you do NOT see in documentaries, to avoid falling in the usual traps.

Throughout this book, a smaller version of the image below will accompany short discussions related to documentaries.

Next time you watch a documentary, pay attention. Notice the tricks used to convey complex messages through simple words and images. Look for features you use in your slide presentation software which they DO NOT use in documentaries. Get inspired!

2 Basics

J.-P. Dionne, *Presentation Skills for Scientists and Engineers*, https://doi.org/10.1007/978-3-030-66069-7_2

2.1 Don't Compete with Yourself! Avoid Text!

When delivering a presentation, you provide two types of info:

- Visual: Your slides
- Verbal: Your voice

Most people are "visual", that is, if they see and hear things at the same time, their focus will be on what they see. This implies that if you display text on the screen, most people will read it. All of it. And if there is a lot of text on the screen, they will read it all.

And they will stop listening to you.

They will even wish you could shut up while they read. It is very annoying to have someone talking when you try reading something. And they will get mad if you move on to the next slide before they are done reading.

The solution is simple: don't have your voice compete with text from your own presentation slides:

- Use as little text as possible: the text should be provided by your voice, not by your slides
- If you do include some text, stick to key words, and make sure you only introduce this text as you talk about it. Not earlier, not later. Just in time, like the Japanese would say (Sect. 2.2)

2.2 Provide Visual Info Exactly When Needed

Think of school lessons provided on old-fashioned blackboards. The teachers talk as they write down on the board. And you can be sure that the teachers provide the visual information at the exact same time they talk about it. Why? Can one really talk about something while writing about something else? On the blackboard, the teacher provides visual info perfectly timed with the auditive info. That's ideal.

Unfortunately, in the vast majority of slide presentations, visual information is provided way ahead of it being discussed by the speaker.

And what happens when visual info is provided ahead of time? The audience starts analyzing the visual info (reading the text, deciphering graphs).

And they stop listening to you… And you bother them by talking while they are busy reading your slide.

Make sure you provide the info exactly when needed. Not before, not after.

- If you show bullet points (in case you don't follow my advice and still show text), make each bullet appear individually, as you introduce it
- Only introduce images as you talk about them
- Use "animations" (see Chap. 3)

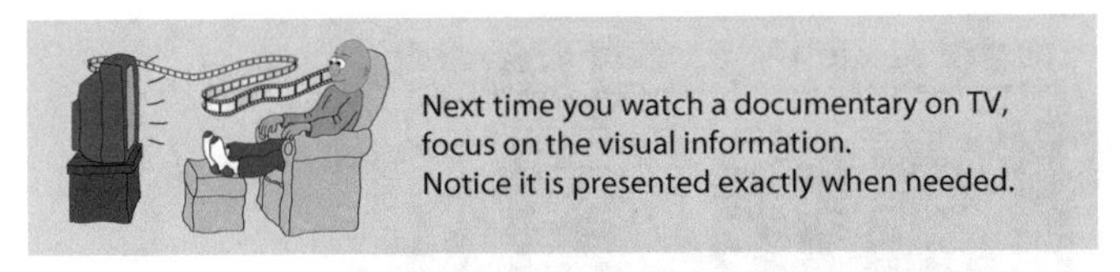

On a blackboard, the verbal and visual info are always in-sync!

2.3 Stick to the Same Fonts

It is so common to see slide presentations filled with text in various fonts (for instance, Times New Roman, Calibri and Arial all on the same slide). Maybe you don't care. Many people don't care. But some do (I do). And those who do care get annoyed by a mishmash of fonts.

My advice: use the same font throughout, or if you vary, do it consistently (for instance, Arial in tables, Calibri elsewhere, etc.)

Those who don't care when you use multiple fonts will not care either if you use fonts in a consistent manner. They will not notice. On the other hand, those who care will appreciate. No one gets hurt, everybody is happy.

Also, only use simple fonts that are easy to decipher. The focus has to remain on your content, not the container. Consider using Arial or similar fonts. Try not using fonts smaller than the equivalent of Arial 18.

In general, try using black colored font (Sect. 2.6), ideally on a white background (Sect. 2.5). The audience hates to have to fight to read text on slides (e.g. yellow on orange, red on blue).

You can use lighter (grey) font to display information which is not as critical (e.g. credits on photos, references).

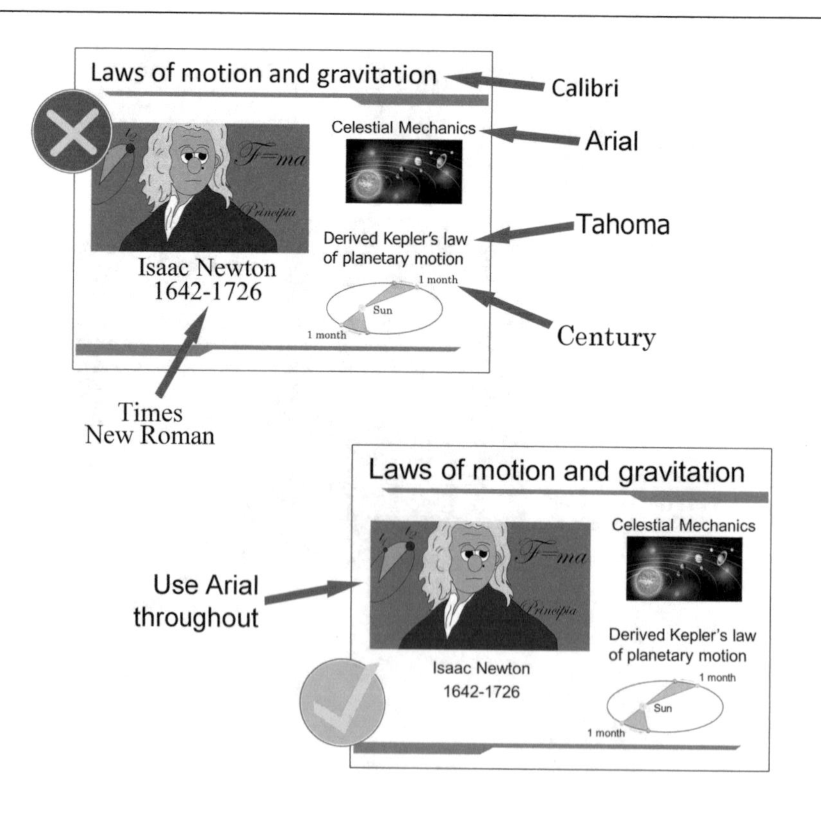

2.4 Leave the Template Aside

Most slide presentations are built using "background" templates, providing continuity across the slides. Such templates are also used to project "corporate" images when one presents on behalf of a specific organization (company or university logos).

But the template should not be a prison. Such templates somehow demand the inclusion of a slide title. While slide titles can indeed be useful and informative, there are occasions where no title is truly required, and the full screen should be made available to display more important things.

You should do without background templates once in a while, to display large graphs or other graphics elements (or, unfortunately, large amounts of text).

In such a case, just introduce a large white rectangle, slightly larger than the screen size. Then, with this rectangle selected, apply the "move to back" function. You can then build a brand-new slide free from any template background.

Do documentaries show a constant "template background"? At most, they include a light transparent logo at the bottom right of the screen. Consider using a full white background with just such a quasi-transparent logo representative of your organization.

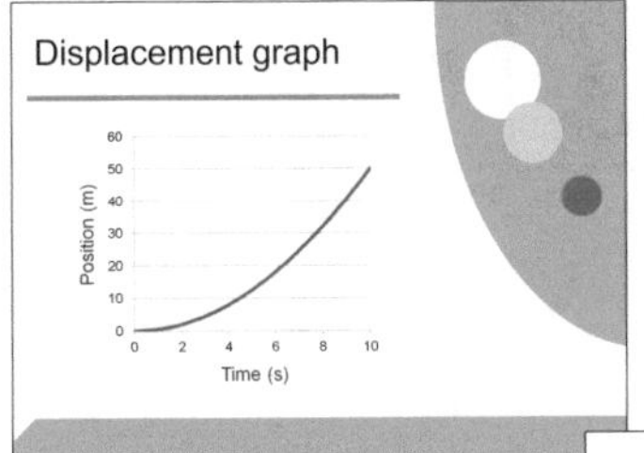

The template restricts the size of the graph by limiting the workable area

The white background from the graph is visible (see Sect. 2.5)

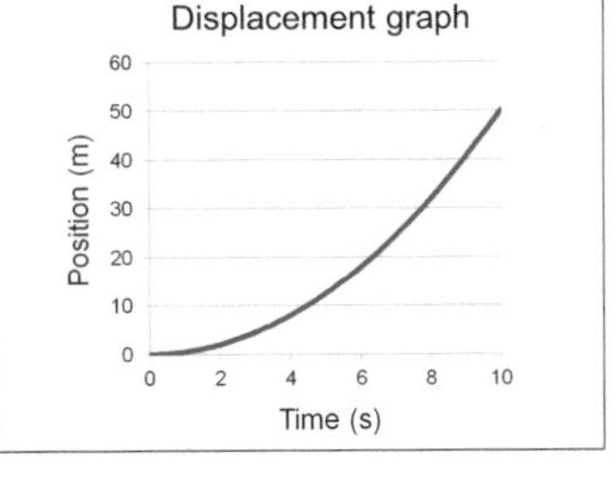

By removing the template, the graph can be made substantially larger

Plus, you do not necessarily need a title for each slide. You can save even more space

2.5 Choose a White Background

As mentioned in Sect. 2.4, you do not need to use a template all the time. But very often, a template will be dictated to you. Hopefully, the template provided to you features a white background. If not, try influencing whoever in your organization provides the template to go for a white background. Or modify it yourself to achieve that goal. And if you have the luxury of selecting your own template, once again, go for a white background. But why?

Most graphs you will create outside of your slide presentation software will feature a white background. Many images downloaded from the Internet also include a white background. And when such graphs and images are placed on a background which is not white, one ends up with a series of undesired white contours.

One way to get rid of these white contours is to 'remove the background' from images (See Sect. 4.8). But while doable, this is not so easy with graphs. Plus, unless you dedicate substantial time removing backgrounds (using Photoshop for instance), the result might not be great, and you might end up with a bunch of white dots or fuzzy white contour around the images. None of these problems occur when using a white background.

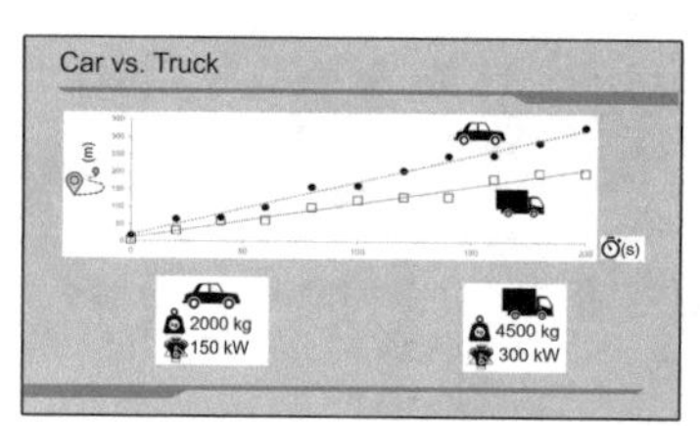

When the template background is not white, you might be left with white contours from graphs and other images, for a poor appearance

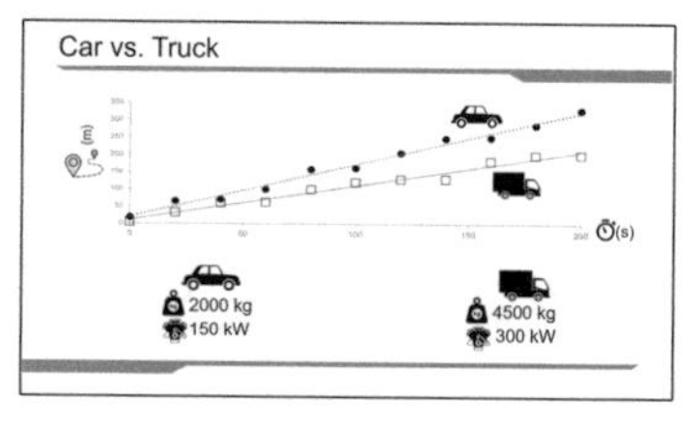

A white background solves these issues

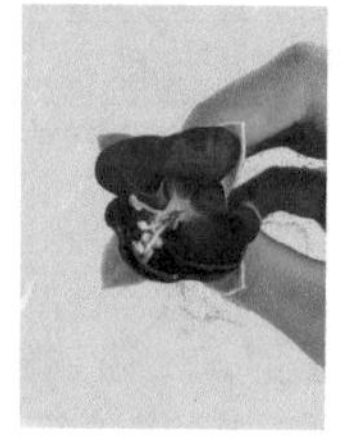

Removing white backgrounds from images to look better on a different color background can prove challenging at times. Image contours might end up fuzzy and white "spots" can remain

2.6 Use Color! But Don't Abuse!

For the first few decades of television, screens were 'black and white', and yet, viewers enjoyed watching movies, shows and the news. Even this book, which uses greyscale for many illustrations, still manages to convey useful information (hopefully). But this being said, color is a great thing, especially for presentations! Colors should be used, for instance, to:

- Distinguish categories of data in graphs (Sects. 6.4, 6.6 and 6.12)
- Better link graph legends to datasets, using text of the same color as data (Sect. 6.6)
- Distinguish values in a table (e.g. negative values in red, positive values in green)
- Highlight good vs. bad results (green vs. red again)
- Emphasize some specific text (warnings, critical results and conclusions)

Images and photographs in colors are of course richer. And colors make slides more appealing, which is critical to maintain interest from your audience.

But one should not abuse color. You should not use color unless color really brings a tangible benefit. For instance, stick to black font, unless colored text really conveys something specific. But use white text on top of dark-colored surfaces.

Just don't use color for the sake of using color. Too much color might just mean too much distraction. Let color be useful… or stick to black and white!

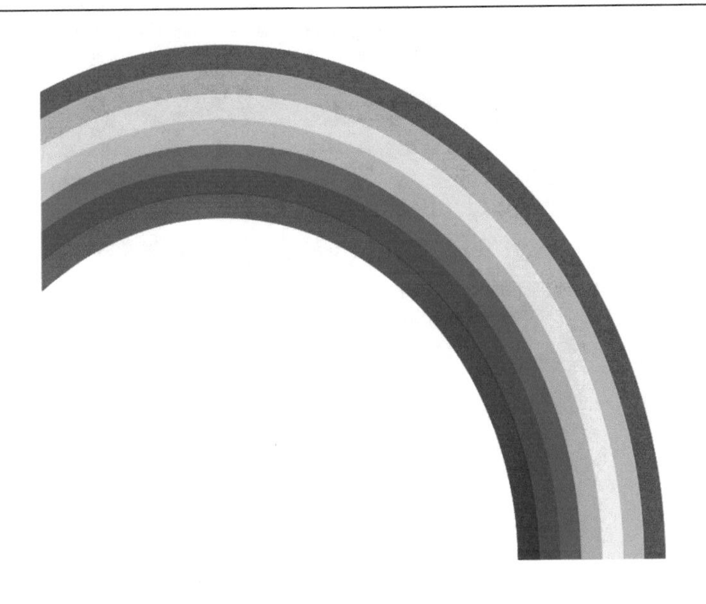

2.7 Choose the Right Screen Ratio

Most projectors use the 16:9 screen ratio. But some people keep using the old 4:3 ratio. As a result, two black columns appear on each side. And one loses an opportunity to use that very useful space on the screen.

Prior to presenting, ask for the screen ratio. If you started with the wrong screen ratio, beware of changes that can be made to your presentation as you transfer it to the appropriate ratio. Many images and text boxes will get distorted. Get ready for having to fix everything. Including possibly the background template!

You can also adjust the screen ratio in line with paper formats (e.g. A4) in case you want to use your slide presentation software to create printed documents!

A 4:3 slide projected on a 16:9 screen displays two ugly large black bars

And you lose the opportunity to use all that precious real estate

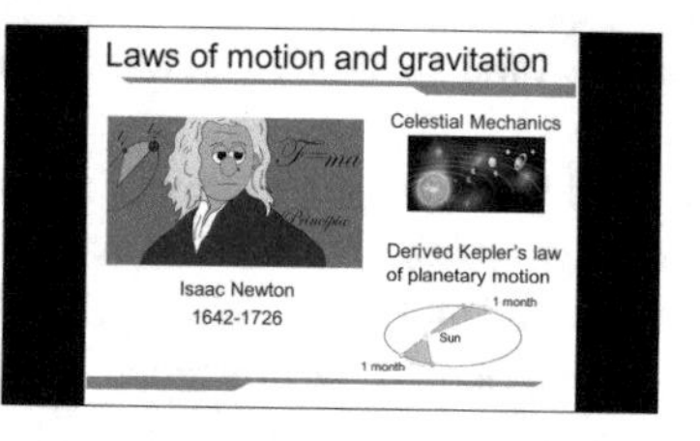

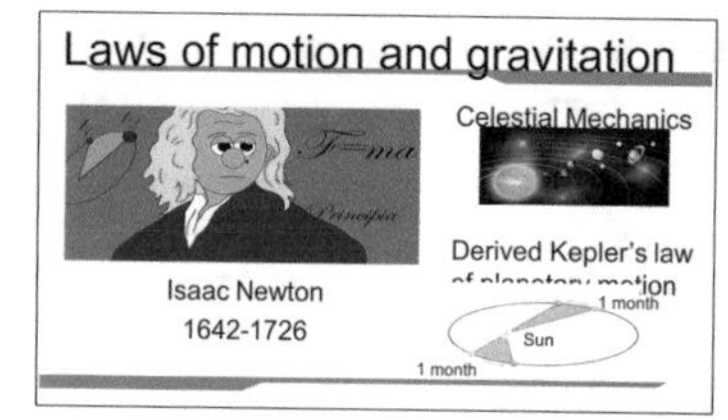

Beware...

When switching an existing 4:3 slide to a 16:9 slide, both images and text get distorted. Some fixing is required!

After some fixes, the slide looks good again in a 16:9 ratio

You can take advantage of the extra space provided

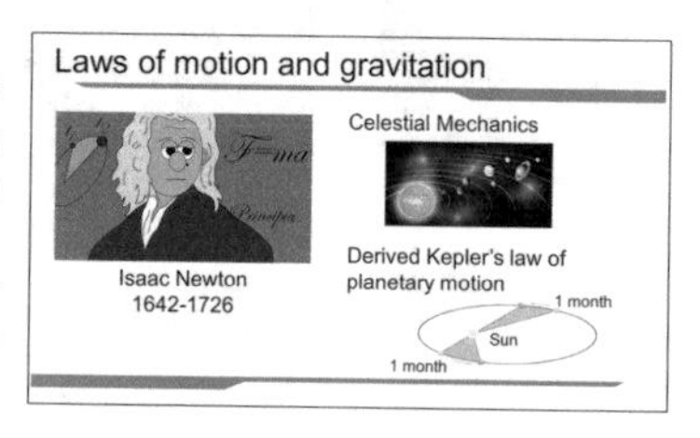

2.8 International Audience?

If you are presenting to an international audience, with a majority of the crowd not speaking English as a first language, you have to be careful about a few things:

- Use less text (this tip actually applies all the time – Sect. 2.1):
  - Don't use full sentences, just fragments (bullet points)
- Use tons of images and icons illustrating your main points:
  - Each text box or bullet point should have its associated image
- Talk slowly and:
  - Avoid slang, avoid idioms, avoid cultural references and jokes
  - When synonyms exist, only use the simplest term
  - Clean up your speech from filling words ("you know…")
  - Be concise and straight to the point

If you follow these tips, it is actually not only the international audience that will benefit. Even the native English speakers are likely to better understand what you are talking about.

2.9 Slides Reviewed for Approval?

Presentations from government employees are typically filled with text. Why? They have to go through layers of approvals, by officials who might not have a clue about the presentation topic. Reviewers would be at a loss with just a bunch of images and graphics. Unfortunately, this results in bad slide presentations.

If you have to deal with such constraints, what could you possibly do to enhance the quality of your presentation?

- Check with your reviewers whether you could provide your text as "speaker notes" within the presentation file. This would then give you full freedom on your presentation content. That would be ideal
- At the very least, minimize the amount of words used. Drop a few words (e.g. articles), avoid repetitions

Finally, the most important point:

- Animate your bullet points! Do not make your text appear all at once on the screen. As mentioned earlier (Sect. 2.2), the audience would just give up on you and start reading the text
  - Make bullet points appear one at a time (easy using slide presentation software), or animate as separate items
  - The reviewers might only review a PDF version, and as such, they might not even notice your animations

---

Providing reviewers with speaker notes might allow you to remove text from the slides

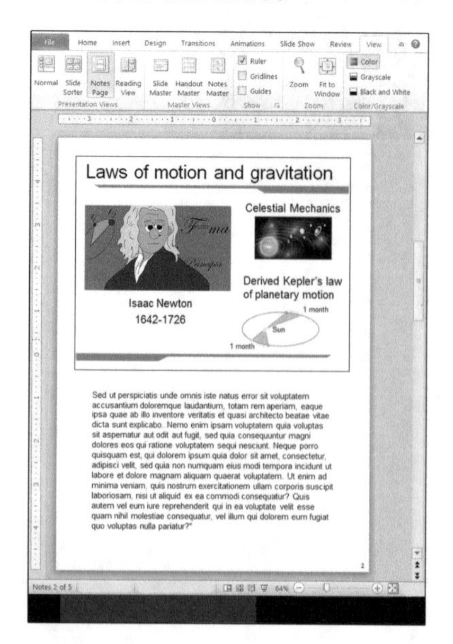

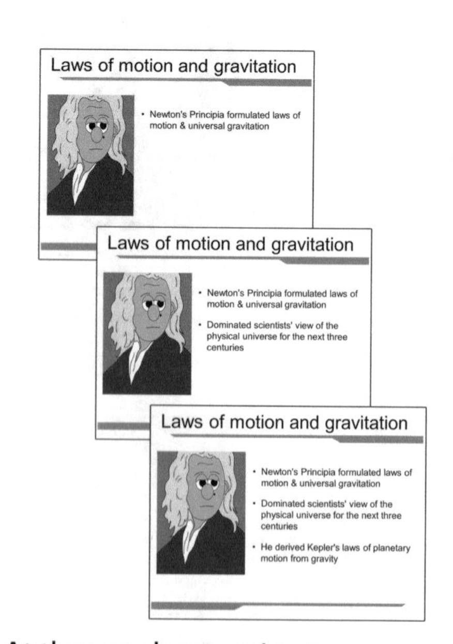

At the very least, animate your bullet points!

Animations

3

J.-P. Dionne, *Presentation Skills for Scientists and Engineers*,
https://doi.org/10.1007/978-3-030-66069-7_3

3.1 Avoid "Fancy" Animations

Animations are extremely important, so that the visual cues appear at the right time, that is, when the speaker starts talking about them (see Sect. 2.2).

But only use two types of animations: "appear" and "disappear". Forget about text and images flying all over the screen and doing wild things. Such craziness just distracts from the important message you try to convey. Appear. Disappear. That's it.

Also, I strongly suggest you get rid of any "transitions" between slides. I find them annoying. Especially when you animate using multiple slides (see Sect. 3.2).

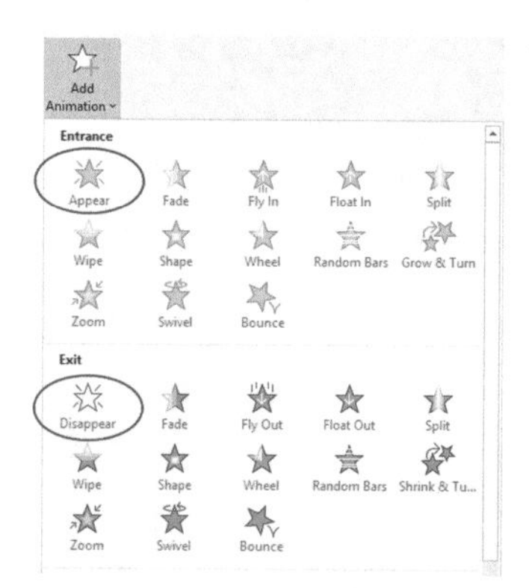

When animating, I strongly suggest you only use "Appear" and "Disappear"

All other options just offer unnecessary distractions

I also suggest you use no transitions between your slides (select "None")

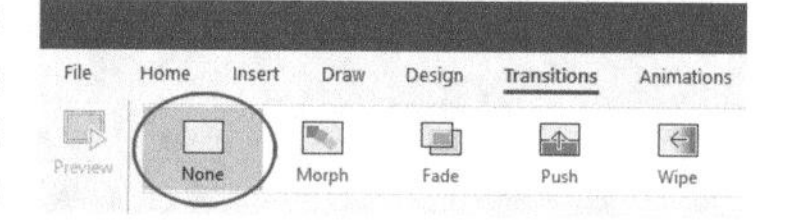

3.2 Option 1: Animate on Multiple Slides

There are two main options to generate animations on your slides. The first one consists in creating a new slide for each element that appears on the screen. For instance, if you first want to show an image, then circle something on the image, and then add text, you could introduce three slides:

- Slide 1: only show the image
- Slide 2: create a copy of Slide 1, and then add the circle
- Slide 3: create a copy of Slide 2, and then add the text

Pros:

- Such "animations" are very easy to generate
- They do not require knowledge of software animation tools

Cons:

- If you made an error or wish to make changes, you might have to apply the same correction on multiple slides
- You end up with a larger number of slides (which can be misleading when trying to estimate the duration of your talk based on the number of slides)
- Your computer file size will be larger

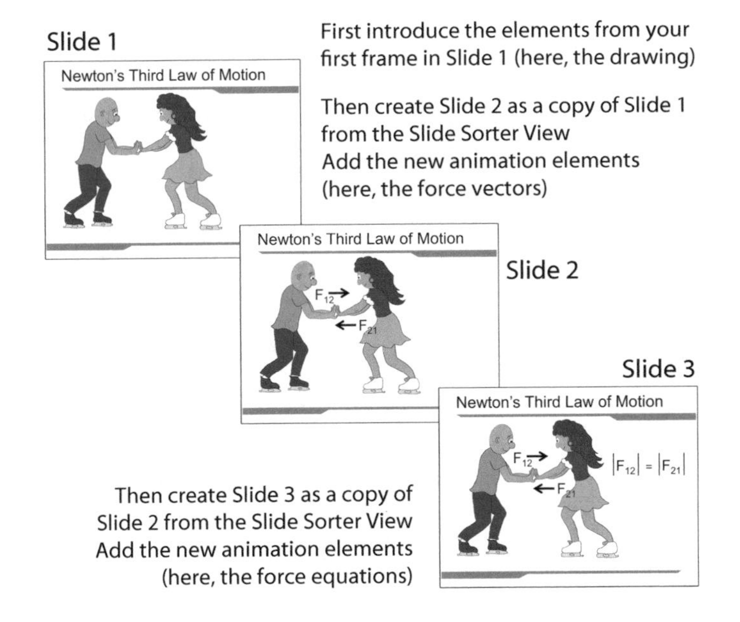

3.3 Option 2: Use Animation Tools

The most elegant option is to use the built-in software animation tools, working from a single slide with all elements eventually shown on screen. You select each individual item to be animated and assign an animation action to it, using the animation menus (only use "Appear" and "Disappear" – Sect. 3.1). You can then modify the sequence of animations or apply other fancy options such as automatic timing, grouping of animations, etc.

Pros:

- Each specific element only needs to be corrected once (as opposed to potentially multiple times, as in Option 1)
- You do not end up with multiple slides to generate what is really one single slide
- You have access to more animation options (e.g. timing, grouping, etc.)
- You end up with a smaller file size, compared to Option 1

Cons:

- Your slide might get cluttered, especially if some items are to appear on top of each other – not all items will be visible on the editing screen
- With lots of animations, you might get lost (especially matching items on screen with the list within the animation menu)
- Alternatively, you can create animations using a combination of both Option 1 and Option 2!

Only one slide

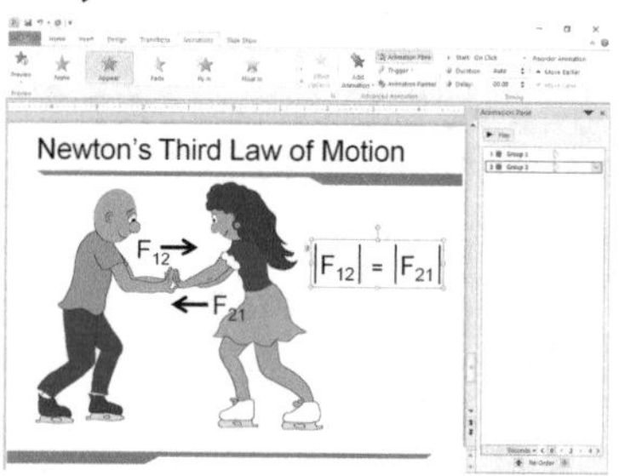

First, introduce all elements from your slide, irrespective of animations

Then, group elements that are meant to appear at the same time (see Sect. 3.4)

Choose the first element to be animated, select "Appear" (see Sect. 3.1)

Choose the second element to be animated, select "Appear"
Keep going until all animations are included

The list of animated items appears on the "Animation Pane", which you can display by clicking on the "Animation" tab, and then hitting the "Animation Pane" button

You can then modify the sequence of animations. Other advanced options are also offered. I suggest you explore the various options. But keep things simple!

3.4 Group to Facilitate Animations

Some elements get animated at the same time. For instance, an image with a caption. To make them appear at the same time, you could animate the image and the caption text individually and separately, selecting the option "appear with previous" for the second element. Or, a simpler and more convenient option would be to "group" the image and the text first, and then only animating that single "group". You can group any elements you want, simply by selecting them (keep the Shift key down to select multiple items), and then choose the option "group".

Pros:

- You end up with fewer animation items in the animation menu
- You are less likely to get lost if you have multiple animations
- You do not have to assign "appear with previous" options for each element
- When modifying the sequence of animations, your grouped items get moved together

Cons:

- You lose some flexibility, which can always be gained back by "ungrouping"

Just be aware that the items automatically created by PowerPoint from templates (e.g. empty text boxes or image placeholders that automatically appear when you insert a new slide) cannot be grouped – you can always delete those and create new ones.

In the example from Sect. 3.3, the two text boxes and the two arrows (4 items in total) are grouped, as shown here

Once grouped, these 4 elements can be animated as a single entity, simplifying the animation process

3.5 Animate Your Bullet Points

I will never emphasize enough: use as little text as possible. But if you do include a list of bullet points, you should animate the bullet points so that they only appear one at a time. Why?

- Visual information should always appear exactly when needed, just ahead of time (Sect. 2.2)
- If you provide text ahead of talking about it, the audience will read your bullet points and stop listening to you
- Too much info at the same time can be overwhelming

To animate a list of bullet points, use built-in software animation tools. Once again, use the simplest possible animation ("appear"). See images below.

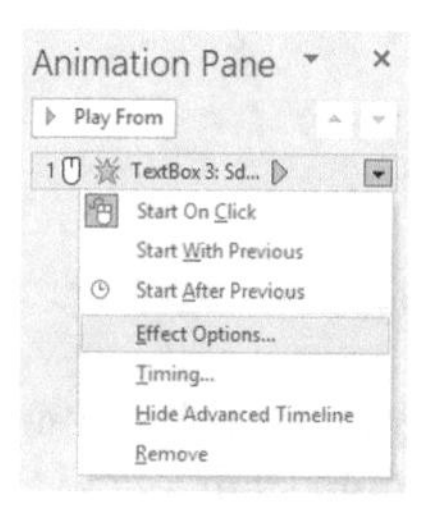

In the example below, select the text box with bullet points

Animate it ("Appear") using the Animation tool

Then, within the animation pane, select "Effect Options"

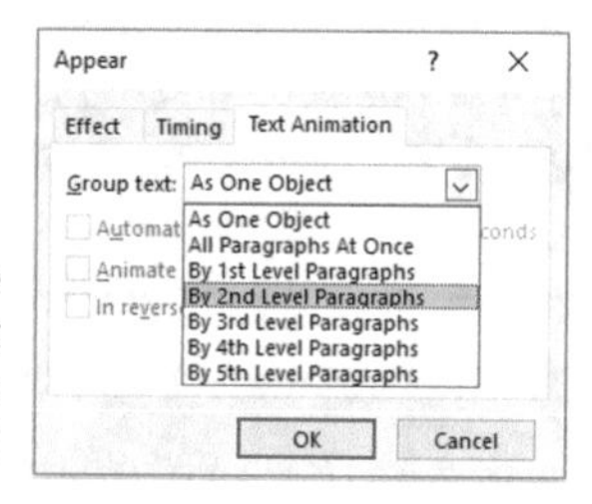

Within the Effects Options, choose "Text Animation", and then select the number of levels of bullets (here, 2nd level)

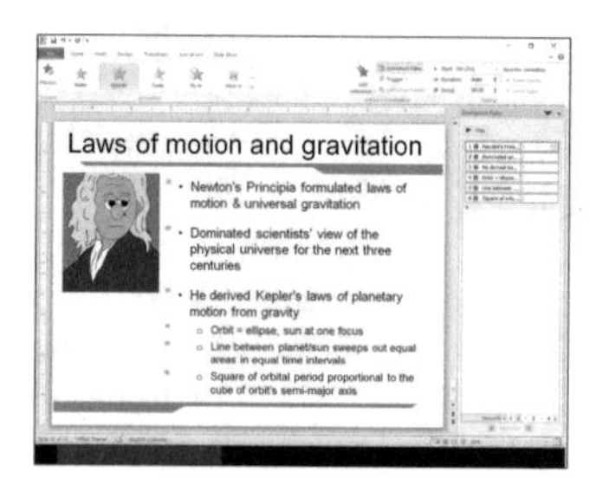

If you expand the item in the animation pane, you even have the option to insert other animations in between each separate bullet (you can have images appear at the same time as specific sub-bullets)

3.6 Apply Exact Positioning

You can move around images, text boxes and graphs on the screen by dragging them with the mouse. You can also align these elements using built-in software align tools.

But also consider making use of the positioning option from the software control boxes to locate elements at precise locations.

This is especially convenient to create animations whereby one element appears, then disappears, then gets replaced by another element which must be at the exact same place.

This is also convenient if you animate graphs (see Sect. 6.10), and you have a few different iterations of a graph that need to appear at the exact same location.

Let's say you want to animate the graph shown below, showing the "45° Throw" curve first, and then the "30° Throw" one after a mouse click

Whether you use Animation Option 1 or 2 (Sect. 3.3 or 3.4), you need the two graphs to have the exact same size and location, so that everything remains the same when the "30° Throw" curve is added

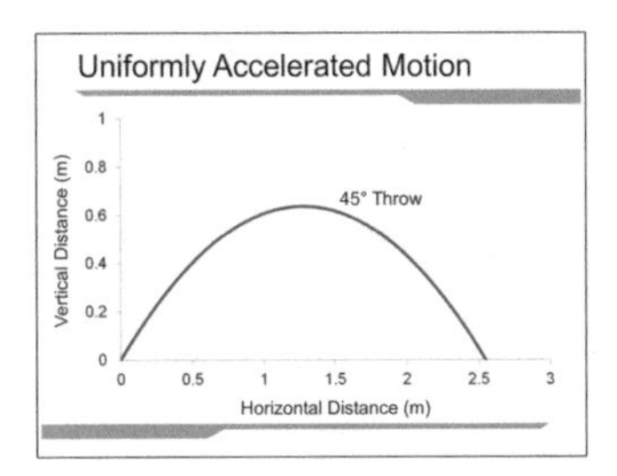

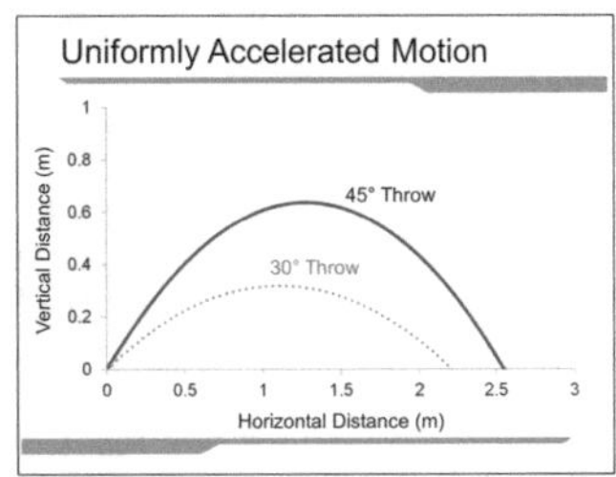

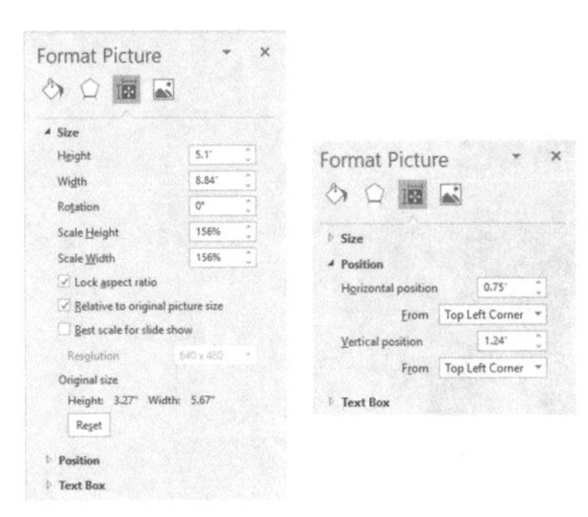

First note down the exact size and position from the first graph (the one with only the "45° Throw" curve), using the "Format Picture" dialog box. The two relevant tabs are shown here

You must then apply the same values (sizes, positions) to the second variation of the graph with both curves shown

3.7 Do You Use Too Many Animations?

You might have heard that too many animations is bad. Indeed, too many "bad" animations is bad. I mean, the crazy animations involving text and images flying all around the screen or appearing in funny ways. Again, avoid those.

Only use "appear" and "disappear", as suggested in Sect. 3.1. If you only use those two, and you apply them so that visual information is provided exactly when needed, you NEVER use too many animations.

Actually, when reviewing your presentation, always ask yourself: could I add MORE animations to ensure visual info is provided exactly when needed? Or to better introduce complex graphs?

Think of documentaries: they use something like 24 images per second!

Why should you restrict yourself to only one image per minute?

3.8 Introduce "Pop-Ups"

In the early PC days, computers were fully dedicated to one single "program" at a time, filling up the entire screen. However, starting with the revolutionary first Macintosh, we all familiarized ourselves with "windows", that is, independently running applications piling up on top of each other on the screen.

Use the same concept to enhance your slide presentations! Let's say that in the midst of introducing a complex graph, you must "open a bracket" so to speak and discuss a concept critical towards the understanding of your data. Consider a pop-up!

Just before you introduce your pop-up, grey-out your entire graph (see Sect. 4.10 on greying out) for better focusing on your pop-up while reminding the audience that you are not done with your graph. Then, introduce the pop-up window somewhere on top of your graph using an opaque rectangle of the appropriate size. As needed, include further animations within your pop-up window. Once you are done with it, remove the pop-up screen, and bring back the original – not greyed out – graph (use either animation options from Sects. 3.2 or 3.3 to achieve this).

Through a pop-up, you have successfully introduced your concept and the audience has not lost track of the presentation flow.

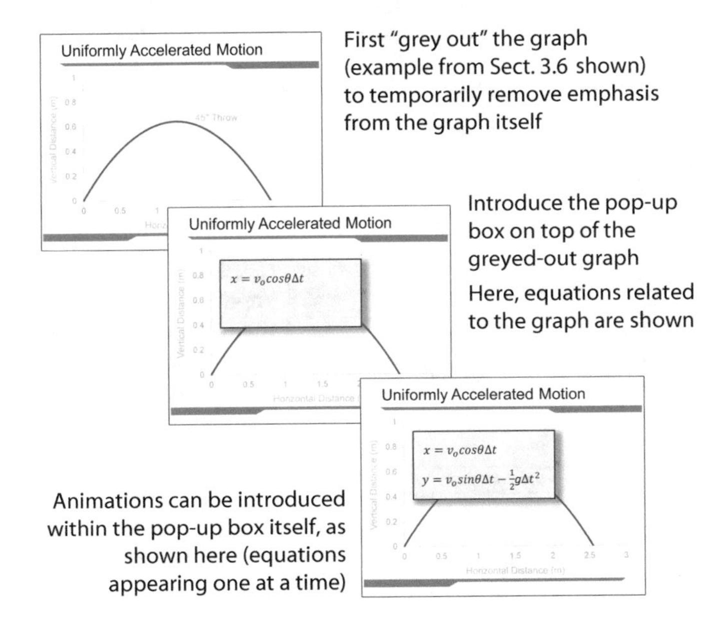

Once this is all done, you can go back to the original (not greyed-out) graph, and resume the presentation

3.9 Maximum Number of Slides Allowed? Use Animations!

Limits on the number of slides are sometimes imposed to ensure that presentations remain short and quick. I find this ridiculous. If there is to be a time limit, it should really be expressed as a time limit, not as a slide number limit!

There exist some rules of thumb (e.g. 1–2 min per slide) which truly apply to slides filled with text (a very bad idea). Effective presentations are likely to include a large number of slides, some of them flashed for just a few seconds, especially if animations are involved (see Option 1 animations, Sect. 3.2).

But if you are restricted to using just a limited number of slides and you can't argue with the conference organizers, then at the very least, "cheat the system". Use a ton of animations! Some of these animations could involve having elements disappearing, leaving room for other elements to appear on the same slide. See the example below.

Would a single image remain on the screen for a full minute in a documentary? No way.

Why should you restrict yourself to the same slide for a full minute for your presentation?

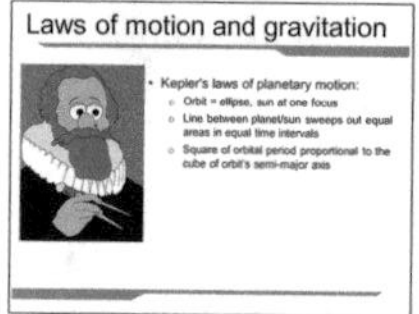

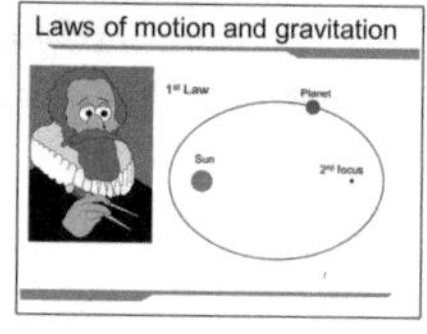

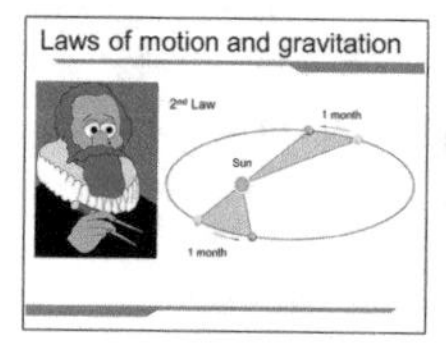

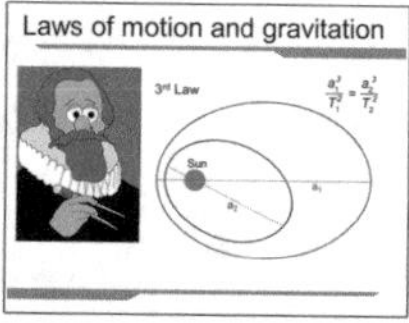

The four slides shown here could be combined into a single slide, with animations

Make the text box from Slide 1 disappear, then Graph from Slide 2 appear, then disappear, then Graph from Slide 3 appear, etc.

(Use Animation Option 2 from Sect. 3.3)

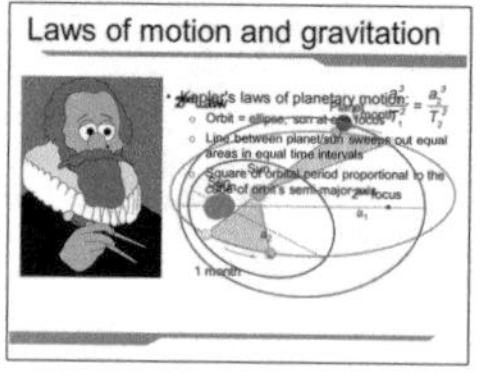

Unfortunately, the resulting slide, in "Edit Mode", looks very messy. But you have cheated the system!

4 Images

J.-P. Dionne, *Presentation Skills for Scientists and Engineers*, https://doi.org/10.1007/978-3-030-66069-7_4

4.1 Use Simple and Clean Photographs

You are likely to require photos for your experimental setup, or images to illustrate your background information, etc. Make sure your photographs are simple, focusing on what you want to convey. If you have the luxury of taking new photographs expressly for your presentation, here is some advice:

- Remove anything not needed in the frame (for instance, empty cardboard boxes next to your experimental setup) prior to taking the photo (cleaning up prior to shooting is much easier than Photoshopping after the fact)
- Ensure there is enough light, but avoid any undesired reflections
- Ensure there are no bystanders, unless they play a role
- Feel free to crop photos to any aspect ratio, feel free to distort them, etc.

Similar principles apply for movies (remove any junk prior to filming, ensure there is enough lighting, etc.) For movies though, it might not be that trivial to crop, so you might have to work with whatever frame you filmed with.

Follow these tips whenever you take photographs or make movies. You might need the resulting images one day for a presentation.

You do not want to show a photo of your test apparatus with that much "crap" in the background

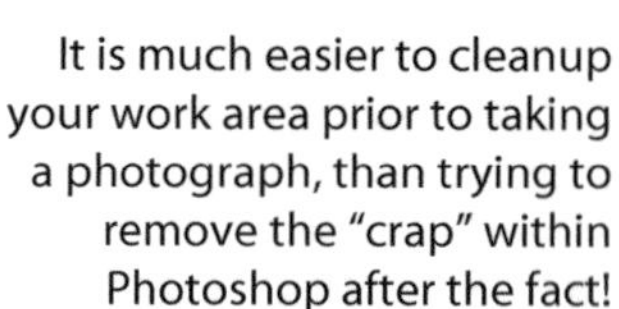

4.2 Avoid Close-Up Photos!

If you are taking a close-up photo of something rather small, resist to the temptation of setting the perfect frame right away, by getting very close to the subject.

Close-up photos, unless you have the appropriate professional zoom lens, is likely to result in images which are only partly in focus (depth of field is very limited) and/or distorted.

Instead, move back, take a picture from a larger frame, and then crop later on from the computer (even within your slide presentation software).

The image files are so large nowadays, that you really don't lose much resolution when cropping photos and enlarging them later to put them on your screen. This way, the image will be totally in focus.

Take photos from a distance, to ensure that the entire frame remains in focus. Just crop to what you need later on, on the computer

4.3 Complement with Internet Images

The Internet is filled with freely accessible images to complement the other illustrations directly originating from your work. Use Internet images to illustrate concepts, to accompany your spoken words, or to make your slides prettier. Images can be very effective at expressing complicated concepts.

Similar to the photographs you take yourself (see Sect. 4.1), ensure extract simple images from the Internet. Images must not distract the audience from the main message of your talk.

- In your search engine, type the concept you are trying to illustrate. You are likely to be provided with illustration ideas you had not thought of
- Use cliparts. Include the word "clipart" in your search key words
- Favor images with white (or transparent) backgrounds, or black and white (for instance, silhouettes, see Sect. 4.4)
- Feel free to modify the images, crop them, etc.

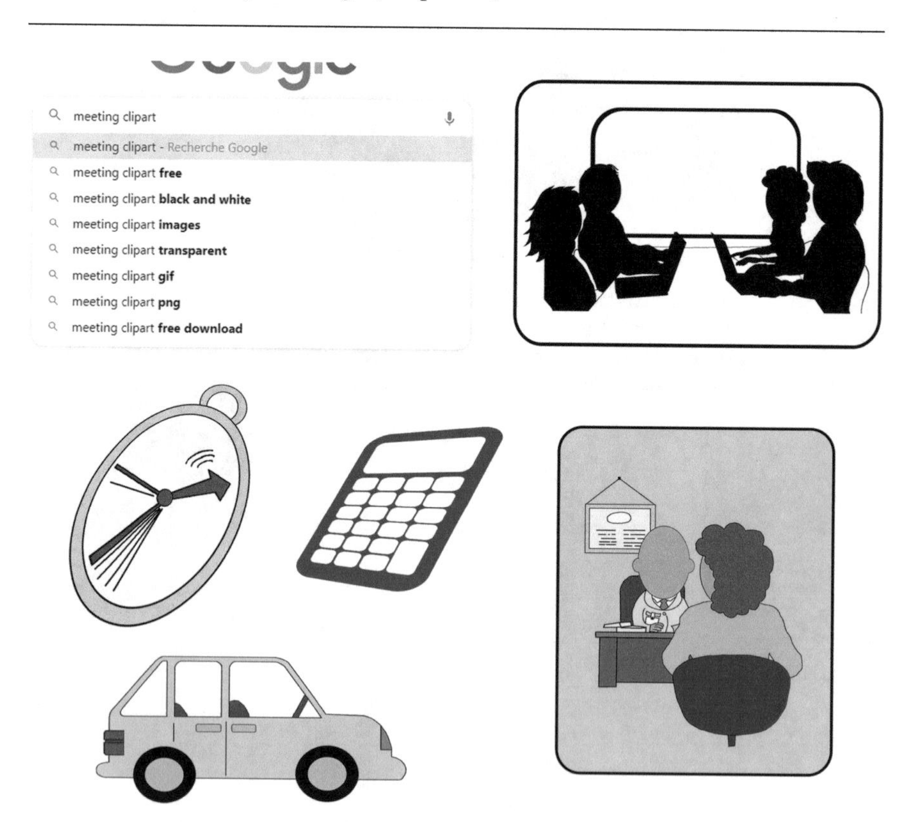

4.4 Generate Your Own Silhouettes

As discussed in Sect. 4.3, cliparts are convenient to illustrate concepts without distracting the audience. The most effective cliparts are called "silhouettes", consisting of black and white simplified drawings (no shades of grey, just black and white). Many of those are available directly from the Internet (add the keyword "silhouette" when searching images).

But you can also create your own silhouettes, starting from either images from the Internet or your own photographs, using Photoshop:

- Crop around the portion of the image you are interested in
- Erase all unnecessary areas, leaving only the item of interest, without any background
- You might need to enhance the brightness and contrast
- Then use the "Threshold" tool from Photoshop, adjusting the threshold level until you get the best effect
- Proceed with a final cleanup. Use a black or a white pencil to make a few final touch-ups to add or remove details

Consider using silhouettes to illustrate your experimental setup, icons for your graphs, etc.

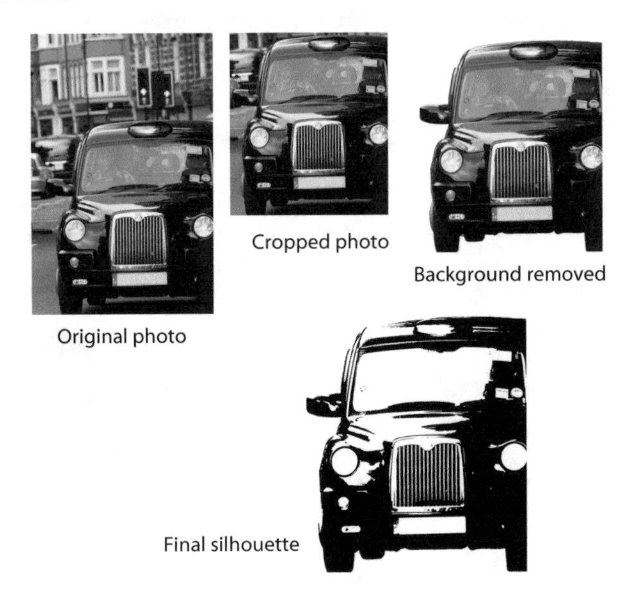

4.5 Use the Screen Capture Button

Anything you see on the screen, whether from the Internet, or from any other application, can be extracted as an image and brought to your presentation. Hitting "Print Screen" is equivalent to a "Copy" action: the content of the screen gets copied to the "clipboard".

The resulting image can be pasted in your presentation file, where you can crop it and enhance it using the built-in image tools. Alternatively, the image can be pasted into advanced image processing software, for more options.

For instance, the "Print Screen" function is useful for importing tables generated in spreadsheet applications. They include table creation tools that are much more powerful and flexible than what you find in slide presentation software. Tables created externally can be imported to your slide deck in different ways (Sect. 7.2), but often times the "Print Screen" option provides nicer images. Make sure you remove gridlines in the background, then zoom in on the table (or graph) so that it fills the screen, prior to hitting "Print Screen". A disadvantage of using screen captures though: you cannot edit the resulting table or graph within your slide presentation software. You have to go back to your spreadsheet application to make changes and then re-import the image.

An advantage of using screen capture is that the image size is always reasonable and adapted to your need for a slide presentation (unless you zoom in too much on a portion of the image). "Screen capturing" is also convenient when web pages do not allow you to use the "Copy" function on specific images.

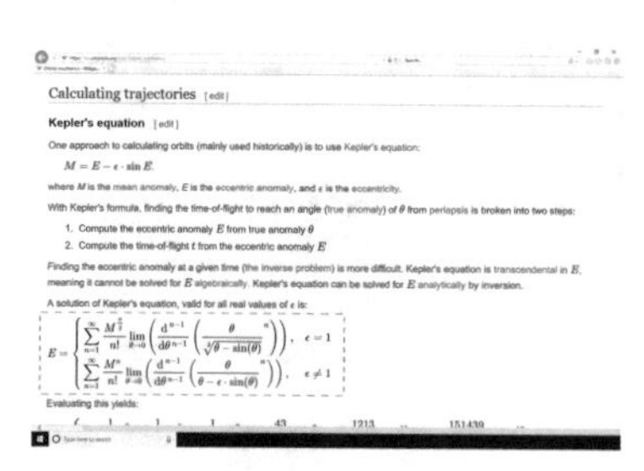

Enlarge the area of interest as much as possible

Then hit the "Print Screen" button and then paste the resulting image to your slide presentation software

Source:
https://en.wikipedia.org/wiki/Orbital_mechanics

By cropping the area highlighted above, you end up with just the portion of the image you are interested in

$$E = \begin{cases} \displaystyle\sum_{n=1}^{\infty} \frac{M^{\frac{n}{3}}}{n!} \lim_{\theta \to 0} \left(\frac{\mathrm{d}^{n-1}}{\mathrm{d}\theta^{n-1}} \left(\frac{\theta}{\sqrt[3]{\theta - \sin(\theta)}}^{n} \right) \right), & \epsilon = 1 \\ \displaystyle\sum_{n=1}^{\infty} \frac{M^{n}}{n!} \lim_{\theta \to 0} \left(\frac{\mathrm{d}^{n-1}}{\mathrm{d}\theta^{n-1}} \left(\frac{\theta}{\theta - \epsilon \cdot \sin(\theta)}^{n} \right) \right), & \epsilon \neq 1 \end{cases}$$

4.6 Use CAD, FEA Software to Generate Illustrations

In addition to the Internet, another potential source for illustrations is specialized software such as CAD (computer-aided design) or even FEA (Finite Element Analysis). You can ask your in-house experts to generate customized illustrations for your presentation, at possibly very little effort on their end.

For instance, if a 3D view of a rectangular box is needed, any CAD tool could generate one. CAD drawings are also convenient to illustrate company products, sometimes more than actual photos.

Finite Element Analysis software like LS DYNA, Autodyn and others, can also generate nice colored illustrations (sometimes through other software like Hypermesh). For instance, LS DYNA includes a "Hybrid III mannequin model", used for car crash studies. Even if you have no interest in car crashes or human body impact, LS DYNA can generate mannequin images in any position, at very little effort from the "programmer". This can be convenient, for instance, to illustrate someone in a kneeling position, at a specific 3D angle.

Think outside the box! Software tools are built for specific applications, but they can also be used to generate nice images for presentations that have nothing to do with their main purpose.

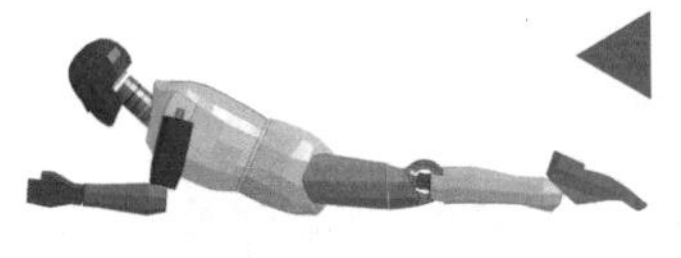

An advanced software (Hypermesh here) is used to generate this image of a "Crash Test Dummy" normally used to conduct tests and numerical simulations for blunt impacts

That image is now slightly modified to be used for a completely different purpose, unrelated to the original intent!

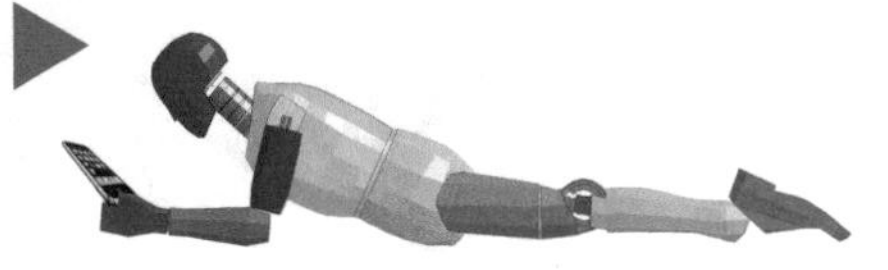

4.7 A Bit of Photoshop...

Photos and illustrations are usually introduced for their educational value. This does not imply they do not have to be pleasing to the eye. Here is some advice:

Cropping: Your own photos or those from the Internet typically come in very specific ratios (for instance 4x6-portrait or 6x4- landscape). For your presentation, there is no need to stick to these ratios. Do not hesitate to crop images, especially if the relevant area of a photo only occupies a portion of the original image, or if distracting info can be excluded by trimming.

Distorting: The space available on your slide might be crowded and a photo might not fit in one of the directions. Do not hesitate to distort it a bit (changing its height without changing its width, or the other way round). No one will notice (similar to adjusting images on a TV).

Enhancing colors: Colors might be "bland". Even within slide presentation software, there exists tools to enhance your images (consider increasing the saturation level and brightness).

4.8 No Photoshop or Equivalent?

There are a few convenient tools within slide presentation software, including:

- Cropping
- Removing the background
- Modifying the brightness and contrast
- Turning to black and white
- Enhancing colors, etc.

Consider enhancing every single image from your slide deck. This is quick and easy with the built-in presentation software tools.

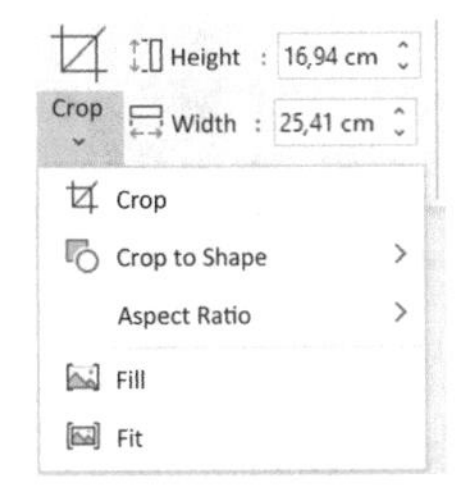

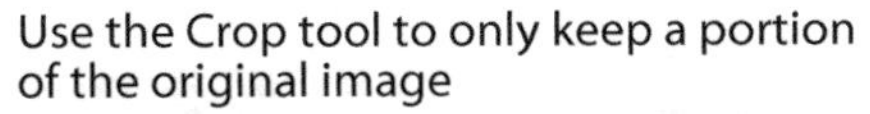

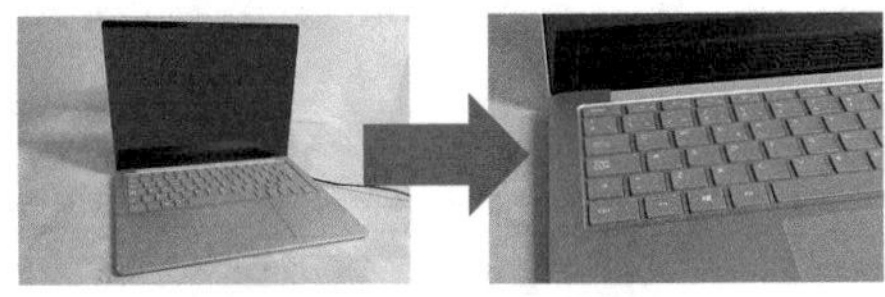

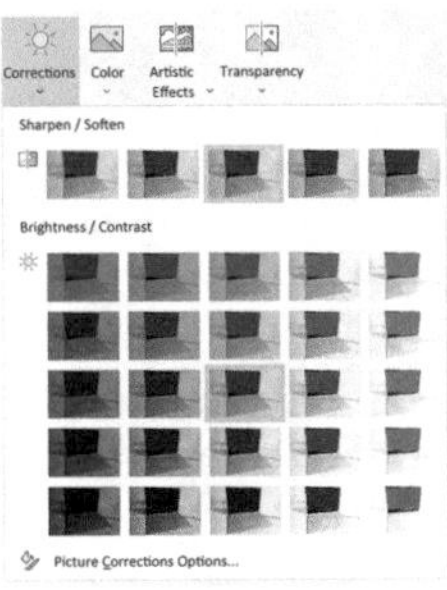

Do not hesitate to use the Image Corrections tool to enhance your images in no time!

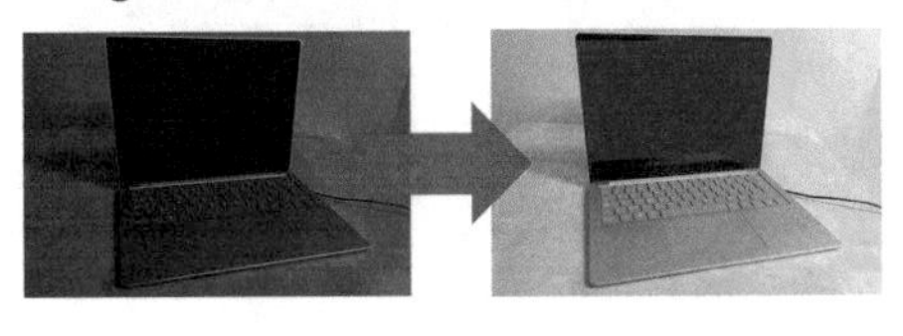

You can even remove the background to end up with cleaner images!

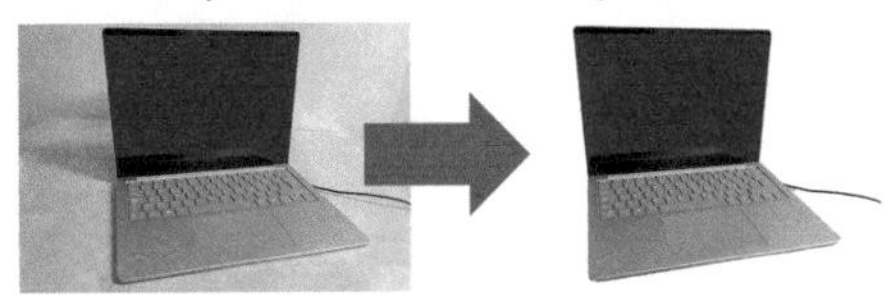

4.9 Layout Images for Perfect Balance

Each slide is like a piece of art. Each slide should look good and feel "balanced". Just like a photograph or a painting, each slide should be perfectly framed. To do so, be consistent in the spacing between the various images or elements from your slide.

This involves some artistic perception. If there is no artist in you, I suggest you focus on maintaining equal or equivalent spaces between all elements from your slide. See the example below.

Talking about art… if you work for a private company, chances are, there are graphics designers within your team. Do not hesitate to reach out to them to get suggestions on how to enhance the look of your slide deck. Little efforts from their end can turn in sizeable improvements. On the other hand, be mindful about potential distractions that could arise from fancy graphics or animations. Keep things simple!

The dotted lines here show that elements of the slide are spaced apart equally, for a nicer visual impression

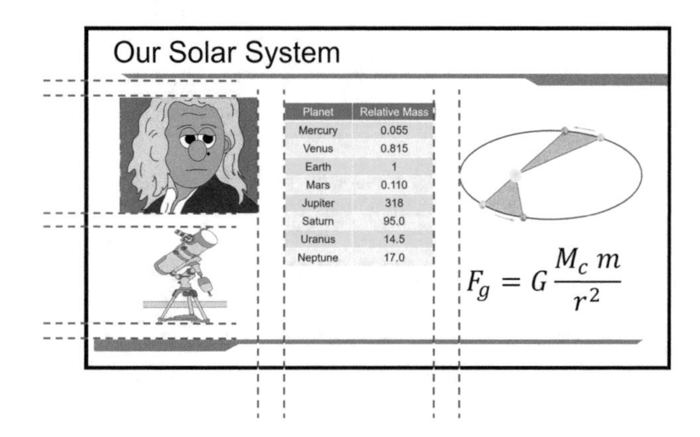

4.10 Grey Out Images to Remove Emphasis

You might have a slide with multiple images or graphics. You might want to emphasize one of them, without necessarily removing the other ones. In such a case, leave the one you want to emphasize as is, but grey out the other ones. This way, the audience does not lose track of where they are within the presentation (at a point in time where all graphs shown have some importance), and they clearly see where their attention should be focused.

To grey out graphs, you could generate an alternative version of the graph using only light shades of grey or turn your graph into an image which you then turn into a light black and white one.

You can also grey out portions of images and photographs, to emphasize the portion of the image which is NOT greyed out (typical example: black and white photo where only a flower is shown with colors).

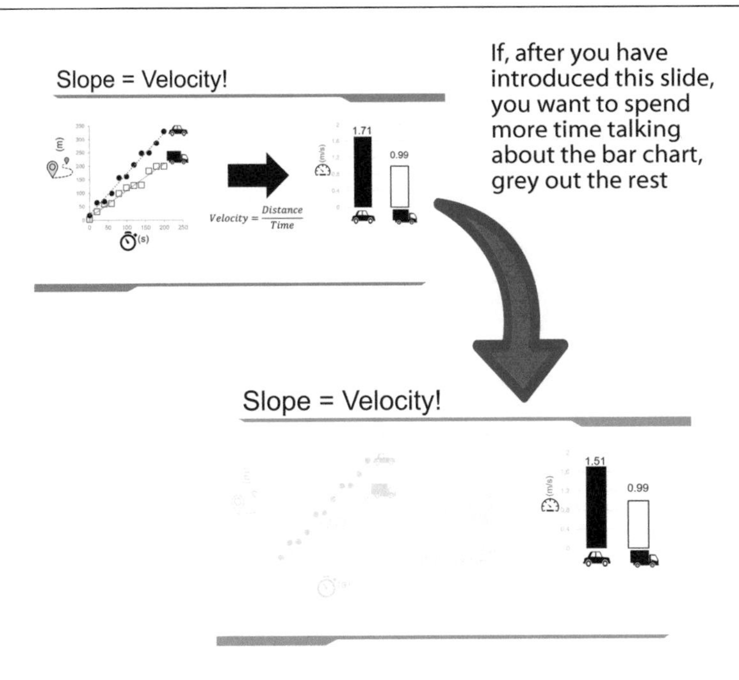

5 Videos

J.-P. Dionne, *Presentation Skills for Scientists and Engineers*,
https://doi.org/10.1007/978-3-030-66069-7_5

5.1 Limit the Number of Movies

Two issues with movies:

- They draw the attention away from you as a presenter
- They never work (not quite but…)

The first issue is not the main one. Really, movies tend to cause the largest number of computer problems, the last thing you want to experience while giving your talk. Movies that work well on your own laptop might not work at all on another one.

Ask yourself whether you really need that movie. But if you really do, here is some advice:

- Also save the movie itself in a directory you can readily locate and run the movie from outside your slide presentation software. But keep in mind that the movie might still fail
- If your movie is a relatively simple one (for instance, a relatively short high speed movie from a scientific experiment), create an animation within a slide, by extracting a couple frames (equally spaced in time), and animate them with an automatic time between frames (Sect. 5.3)
- Or… just be ready to proceed with your presentation without your movie. Have a plan-B

Sound in movies can also be a concern. The hardware might not provide for any sound. Or the sound might be too dim, or distorted. If possible, avoid sounds (see Sect. 5.6).

5.2 Automatic or Manual Activation?

If you know you will run a video at the moment you get to its slide, make it run automatically. On the other hand, if you need to control the exact moment to get it started (for instance, if you also introduce other elements on the same slide), use manual activation (clicking with the mouse).

The automatic activation has the advantage of not depending on the mouse, which can prove tricky with a nervous speaker. You can also consider delayed automatic activation, in which case, a bit of practice can help determining what would be the best delay.

Also, if your movie is short, consider having it run indefinitely in a loop, which can help when a single viewing is not enough. Be careful though: if you talk while you show the video, the audience will focus on the images, more than on your voice. Unless what you say is directly related to what is going on in real-time on the video, chances are they will stop listening to you.

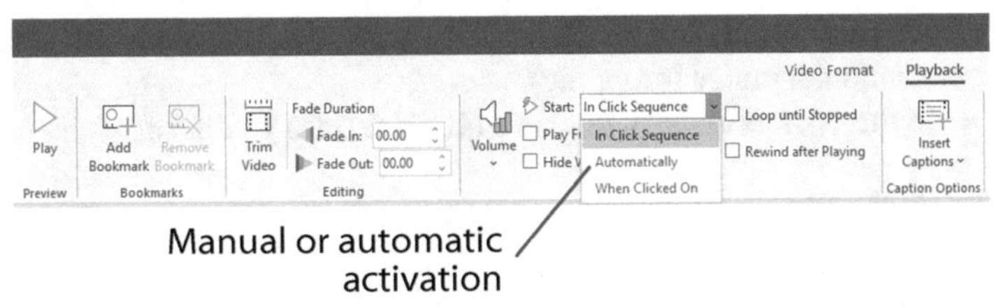

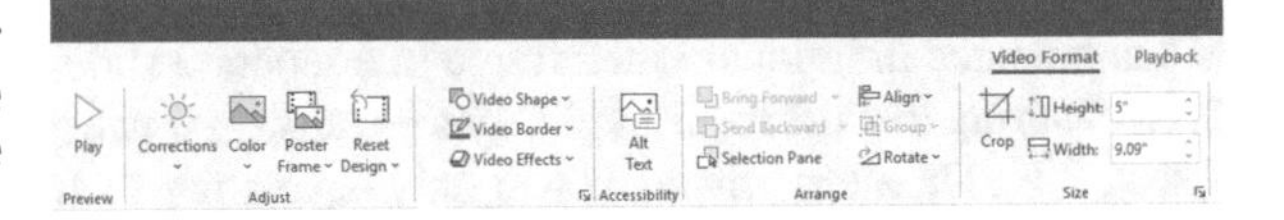

5.3 Replace Videos by Series of Still Images

The safest option, when planning on showing videos, is to replace them by a series of still images which you can animate yourself within your slide presentation software.

Of course, this will not be practical for movies that last more than a few seconds, or when sound is important, or when smooth flow of images is critical. But if you are dealing with a short silent movie (for instance a high-speed movie from an experimental trial) or if you display computer-generated animations, replacing videos by a series of still images is either a very good Plan-B, or can actually replace the video altogether as your Plan-A.

There exist a few options to extract still images from a video. One of them is to use an advanced video editing software which includes this feature (follow the software-specific instructions to turn videos into a series of images).

Another option is simply to rely on screen captures of video frames:

- Play the video in full screen mode using any available video software
- You might press "pause" at the exact frame you want to extract (optional)
- Hit the "Print Screen" button (Sect. 4.5) – paste the image in your slide, remove areas around the image (cropping)
  - Redo the operation until you have all the frames you need

The number of frames you need really depends on what you want to show, and how fast the images change over time. In some cases, 5 frames might be sufficient, while in other cases, many more frames might be needed (which can make this technique not so practical).

You can put each individual video frame in a separate slide (Option 1 animation, Sect. 3.2), ensuring that the resulting images are always positioned at the exact same place (Sect. 3.6). Or, alternatively, if you only have a few frames, you could superimpose them on a single slide, and use the software animation tools (Option 2 animation, Sect. 3.3).

Advantages:
- Avoid the very frequent potential technical issues with videos
- Play the "video" at the pace you want, easily stopping whenever you want, and easily back tracking
- Set the animations to be automatic (e.g. an image per second)
- Include markers, text, arrows, etc. on any single frame
- Show frames that are not equally spaced in time (flexibility!)
- Individually correct resulting images (removing unwanted elements from the frames, enhancing image quality, etc.)
- Zoom in easily
- Save memory space

Disadvantages:

- More time consuming than inserting a video in your presentation
- The resulting "movie" does not flow smoothly
- You lose sound (maybe not such a bad thing – see Sect. 5.6)

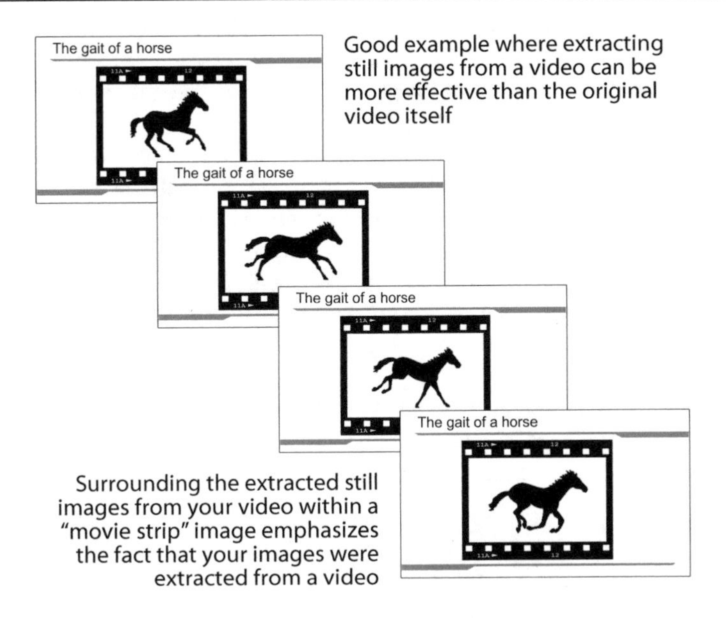

5.4 Generate Your Own Short Videos/Animations

To help with introducing concepts, consider generating a series of still images that run like a video (with proper editing software, series of images can actually be turned into a video).

Consider this simple example: in mechanics (physics), one learns that "the change in velocity corresponds to the area under the acceleration curve". No need to understand physics to illustrate the point here. Just consider that we want to highlight the area under some trace on a graph.

One could simply use color shading: show the original graph (without shading), and then replace the original graph with an identical one with the shaded area.

But here is an even more effective trick: create a series of still images (maybe 4 or 5), with increasing amount of shading, as shown below (in Photoshop, or directly within Excel, using Excel drawing tools), keeping all graphs identical in size.

Animate them in your slides, selecting a fraction of a second between each frame (pick what feels best). This results in a "filling up" effect that effectively highlights the area under the curve.

There are many presentation scenarios where such simple custom-built animations can facilitate the introduction of concepts. Just use the trick!

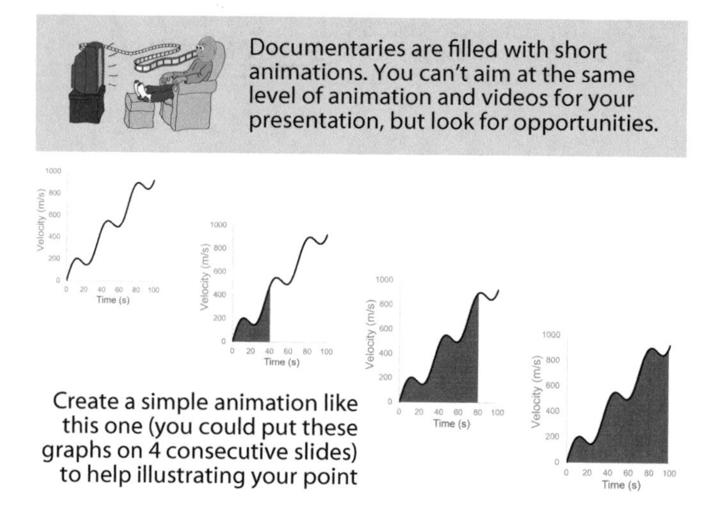

5.5 Your Video Does Not Work?

In almost every conference I attend, there is at least one instance of a video not playing.

There are steps you can take ahead of the presentation to avoid this problem:

- Prepare alternative slides with different video formats (.wmv, .avi, .mov). Most basic video editing software save to different video formats. One of them is likely to work
- Turn video into series of still images (Sect. 5.3), which you insert at the end of the presentation
- Prepare alternative slides without the video, that bring up the main points the video was to convey
- Do a dry run with the audio-visual technician, ideally the day before (Sect. 12.2)

In all cases where you add back up slides, use action buttons (Sect. 10.1) linking your video slides to your alternative slides. Include another action button on your last backup slide to take you back where you started.

Another solution is to leave the slide presentation software, and then run your video directly from the operating system, using any basic video playback software. Have your video saved in multiple formats, to increase the odds of having one version that will work.

If your video fails to work, keep calm! You have many solutions available!

5.6 Should You Use Sounds?

Typically, conference presentations do not include sounds. After all, it is the speaker who is expected to provide the soundtrack!

One exception is when videos with sound are played. Unfortunately, many (if not most) conference rooms are not setup to play sounds. As such, videos are likely to be muted. Even when sounds can be heard, the volume might be too low, or only the front rows might hear properly, causing frustrations.

If the soundtrack is very critical to you, be ready to generate the sounds yourself, live in the microphone! (I have seen that once).

Sounds can also be used in other scenarios, as slide presentation software contains options to play sounds (in the animation tools). You could use sounds to introduce some features from your presentation or serve as additional clues related to specific concepts. You could even ring a "bell" when moving up slides, just like on old Disney tapes to remind kids to turn the page!

But quite frankly, I do not see much value in adding sounds, given the potential technical problems and the extra work required to setup sounds in your presentation.

But if you insist on including sounds, you will at least demonstrate some originality!

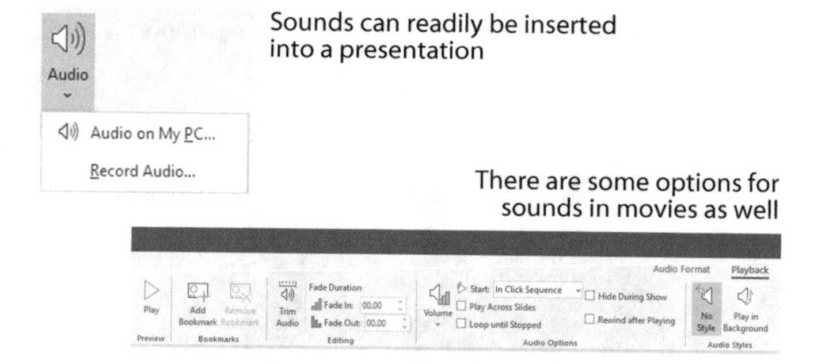

6 Graphs

J.-P. Dionne, *Presentation Skills for Scientists and Engineers*,
https://doi.org/10.1007/978-3-030-66069-7_6

6.1 Invest Time on Your Graphs

Graphs are essential components of most scientific presentations, as well as in many other fields (economics, administration, social studies, etc.) They require a lot of attention to ensure they are optimized for your presentation. As such, the following few sections include a lot of different tips on how to introduce graphs in your presentation.

Please keep in mind:

- Understanding a graph takes time – even for smart people, even for those familiar with your topic, even for smart people familiar with your topic!
- If you start discussing key findings from your graph while the audience is still trying to decipher the basics of it (for instance, the axes, units, legend, etc.) you have already lost them

When you show a graph, expect a few things from the audience:

- They will try to understand your graph – and as they do this, they will momentarily stop listening to you
- They will look at everything on your graph (axes, legend, numbers, trends, colors, any text on it)
  - They will attempt reading stuff in small font
  - They will pay attention to the useless stuff on your graph

Some overall advice:

- Do not just blindly use graphs created for a report, or borrowed from somewhere else
- A graph that is perfectly suitable for a report or a scientific paper is generally NOT suited for an on-screen presentation
  - Large enough for a test report is not large enough for a presentation
- Keep any text on your graph to a MINIMUM
- Anything worth being shown must be large enough to be read
- Anything NOT important must NOT make it to the screen

Think of any documentary where graphs were shown. Notice how simple they were, both in terms of graphics, content, and amount of info presented.

6.2 A Graph for Each Key Finding

When delivering a presentation, you can only afford addressing a few critical findings and key concepts (despite your temptation to cover everything about your work). Think of your summary and conclusion slides: make sure you introduce a graph dedicated to each conclusion point, specifically and exclusively addressing that single conclusion point.

As such, each graph you present must be focused and specifically designed towards presenting one key finding at a time (or maybe two at most). Do not just use one of your raw generic graphs containing more data than you need to share to prove your specific point.

Think of the few critical and important messages you want to share. Build a customized graph for each of these key concepts.

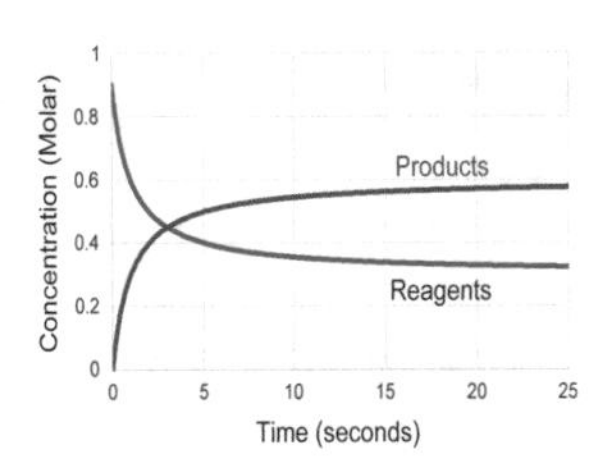

Let's say that one of your important conclusions is related to chemical concentrations at equilibrium.

Based on this graph, one would have to look at the last two points on each curve corresponding to the end of the experiment to determine equilibrium concentrations.

Everything else is a distraction.

A more direct illustration of that conclusion is to extract only these two final data points, and plot them on a simple bar chart

This bar chart focuses on one single fact only, the precise one you want to highlight as a conclusion (concentrations at equilibrium)

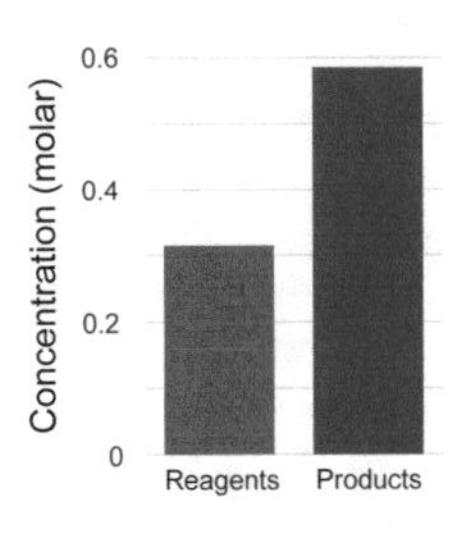

6.3 Favor Bar Charts vs. Scatter Plots

In typical presentations, one typically does not have enough time to provide all details and proofs, especially when sharing experimental data.

Your raw experimental data might consist of thousands of data points that can only be fully illustrated on tons of complicated scatter plots. In addition, your data might contain statistical information (standard deviations, variations, etc.) critical for a stand-alone scientific work.

But for on-screen presentations, generate simple graphs focusing on just one key observation (see Sect. 6.2). For this purpose, histograms and bar charts are ideal. Such graphs present only a limited amount of data, for instance averages of results. The illustration will be obvious and straight to the point, especially if direct comparisons are displayed.

Granted, bar charts typically lack all the detailed statistical information (individual data points, standard deviations, etc.). But the audience must trust that you have done your background work and that the detailed information with all the raw data is available in a test report or a scientific article, or that you can discuss it after the talk. You can emphasize that point verbally.

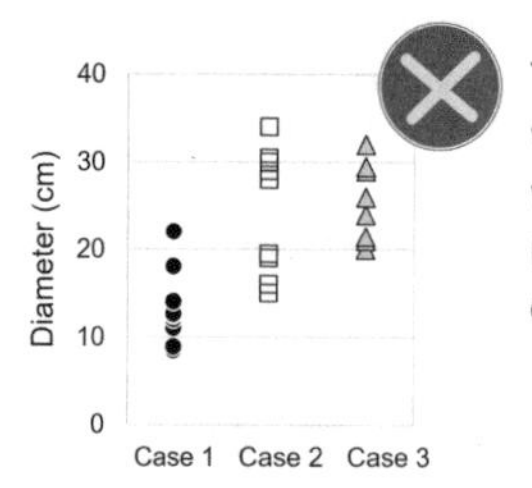

The graph on the left contains very detailed information which is appropriate for a test report, but not so much for an on-screen presentation

Instead, use a bar chart focusing on the main point you are trying to achieve (comparison between the three cases), thereby avoiding distraction due to excess of information

The audience will take for granted that more detailed information can be found in your paper

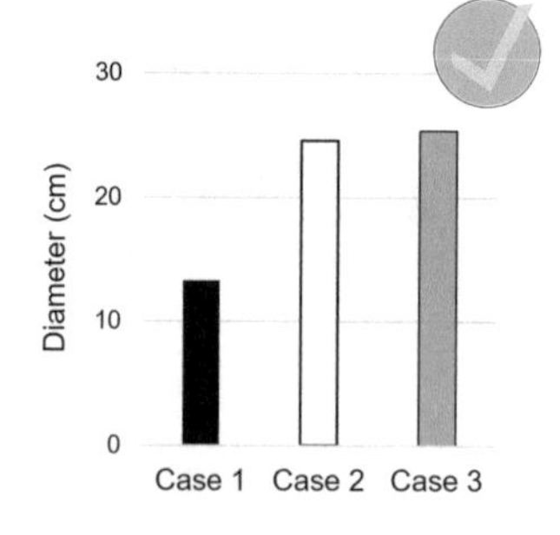

6.4 Make Data Points Clearly Visible

Use LARGE markers to display data on graphs for on-screen presentations, much larger than what you would use for a written report. As long as you can still readily distinguish one data point from another, you can enlarge them.

Similarly, use thick lines to show trendlines and curves.

Do use different colors for different data sets (see Sect. 2.6), but do not only rely on colors to distinguish your different types of data. Using different shapes is good.

Ideally, also use animations (see Sect. 6.10).

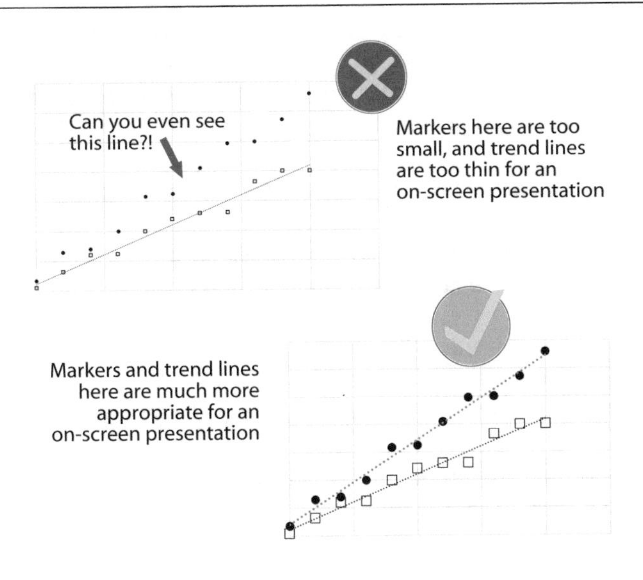

6.5 Use Minimalist Axes and Grids

Avoid showing grids, as they clutter your graphs:

- Grids are useful to extract values from graphs – not something expected from the audience during your talk
- If you insist on including grids, only show the main ones, and use a light grey

Use a minimal amount of numbers on your axes:

- With a graph going from 0 to 100, consider having ticks only for 0, 25, 50 and 100. Maybe 0, 20, 40… But NOT 0, 10, 20…
- Do not show "subticks" – those are only useful on printed documents to accurately extract data from graphs

For graphs aimed at showing trends (not focusing on actual values), use a light grey color for the numbers.

The numbers and titles used for your axes must be in large enough font (at least 16 for Arial), even if actual numbers are not that important

- Limit the number of words as much as possible
- Do not forget units
- Use the same font sizes for both axes. Text can be larger than numbers

Anything you show, no matter how important, must be easy to read and decipher. If really not important, then just don't show it!

This graph is way too "busy"

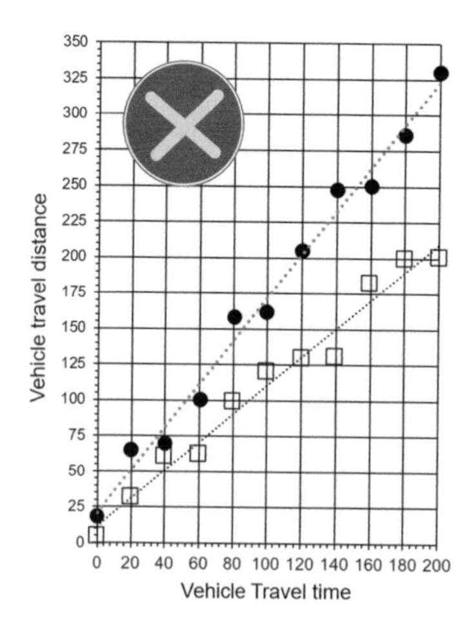

This one looks much better!

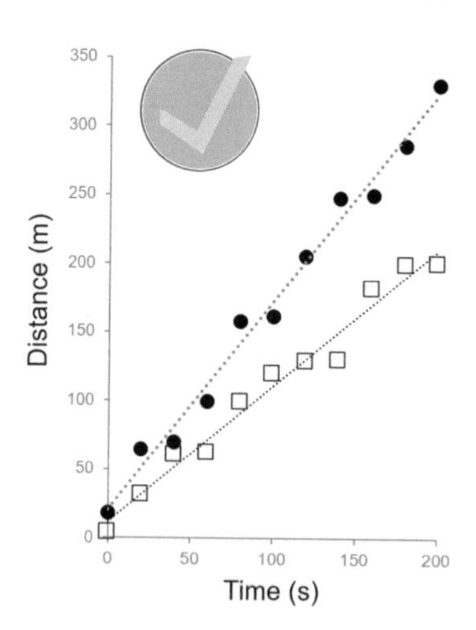

6.6 Apply Legends Directly on Data

In formal written reports, readers have plenty of time to analyze graphs. In such cases, legends can be totally separated from the data set (see first example below). This is not such a bad thing for on-screen slide presentations either, especially if the dataset is animated (see Sect. 6.10).

However, in slide presentations, it is easier for the audience to associate a specific data set with its meaning when the elements from the legend are displayed right next to the data (see second example). This avoids back and forth search for information.

Remember that it takes time for the audience to understand a graph. Give them all the chances, introduce your information slowly and gradually. Take them by the hand.

On a scatter plot, it is much easier to read the legend when directly on data, especially in this case here, where "Truck" appears on top of "Car" in the legend, whereas it is the opposite for the actual data

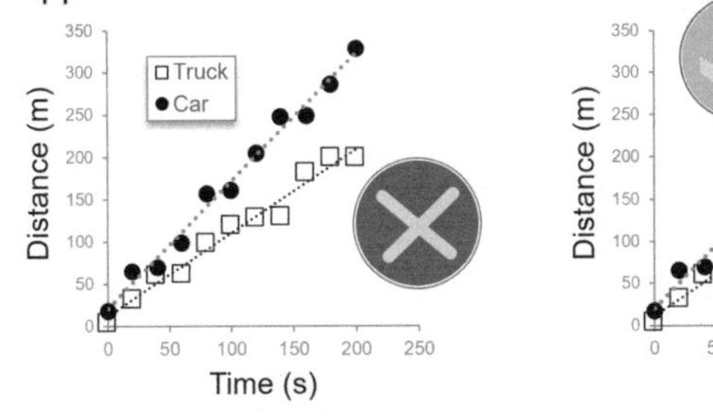

The same principle applies to bar charts and any other type of graph. It is much easier for the audience to quickly grasp which data points relates to which legend category when the legend is directly on the data

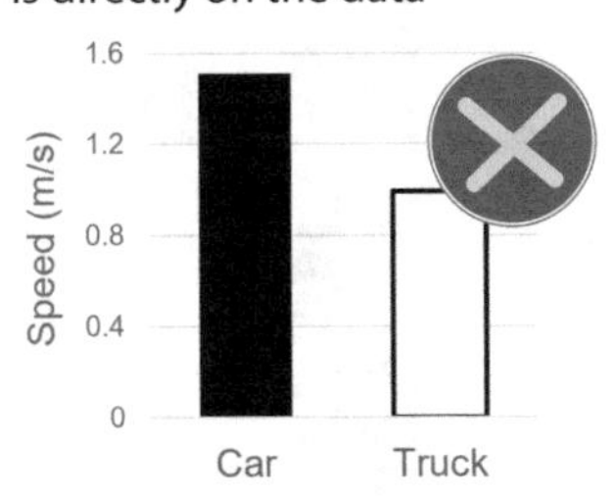

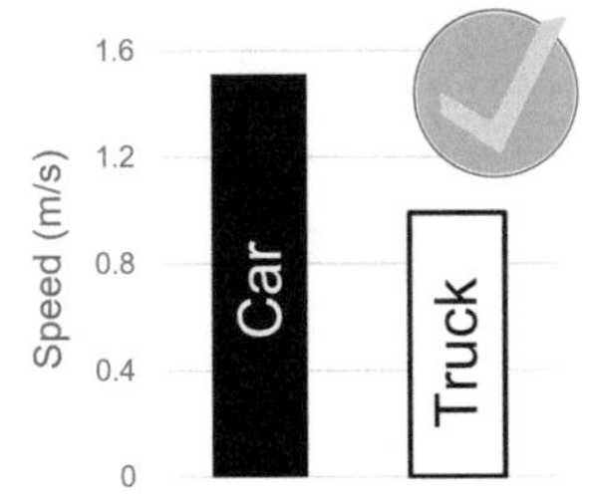

6.7 Use Icons and Images on Graphs

A basic graph only contains the text on the axes, the numbers on the axes, the data points, trend lines, and possibly some text. But to assist the audience with a better and quicker understanding of your graph, do not hesitate to use icons and images, which can also be animated (see Sect. 6.10).

Examples:

- If one axis is "time", you can show a clock icon
- If one axis is "distance", you can use a measuring tape icon
- If one set of data is for "temperature", show a thermometer next to the data
- If one set of data corresponds to a "car", show a car icon

Use very simple images as icons, ideally black and white silhouettes (see Sect. 4.4).

Consider replacing all text with images and icons, so long as the meaning remains obvious (this also applies to legends). An image is worth a thousand words!

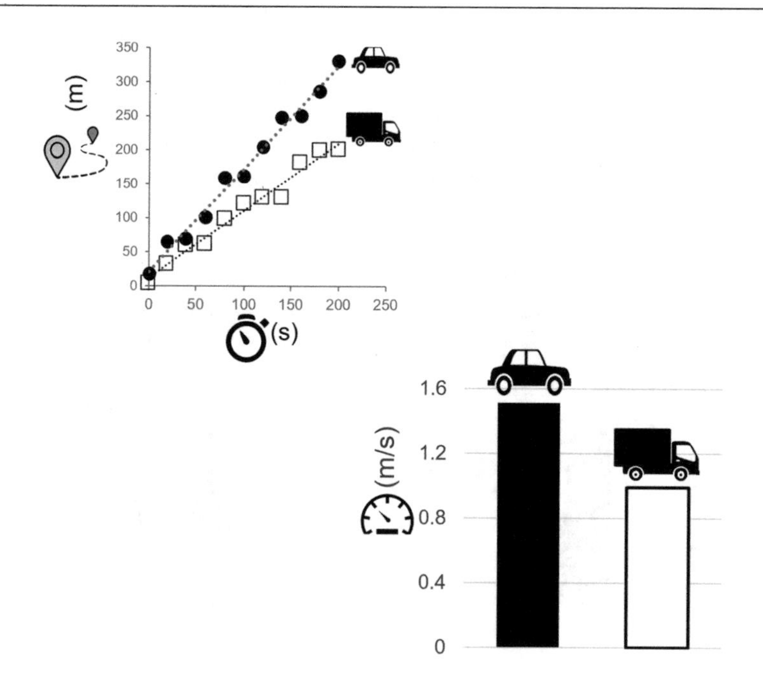

6.8 Add Values Directly on Graphs

If the actual values are of importance, consider displaying the values corresponding to each bar (either on top of the bar, or inside the bar).

But if the actual values are NOT of importance, just do NOT emphasize them.

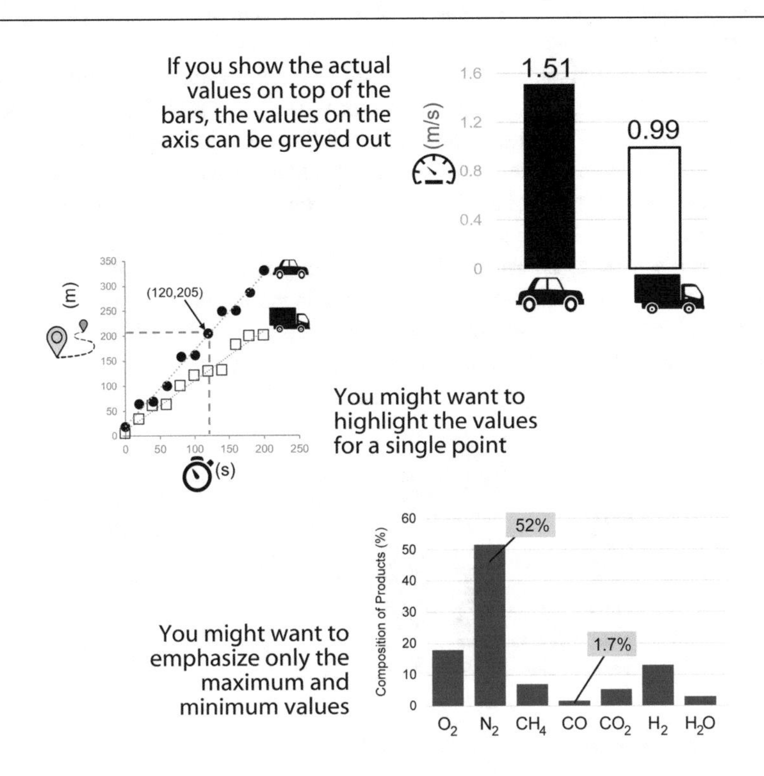

6.9 Beware When Zooming Bar Charts

Graphs are a great way to quickly convey quantitative information. However, in some cases, they might introduce unwanted confusion. For instance, if you compare categories of data in a bar chart, showing values between let's say 425 and 550, it is not appropriate to "zoom" the graph to have the vertical axis start anywhere other than zero (see top figure). In the case shown, while the tensile strength values are well represented graphically, the bar chart gives the false impression that the value for Brass (~500 MPa here) is three times as large as the strength for Be (~450 MPa), which is obviously not the case. Instead, have the vertical axis start at zero (see bottom figure). This way, the relative tensile strengths for the four materials remain obvious (they are actually quite similar here, contrary to what the first bar chart seemed to suggest at first glance).

If you really need to zoom in a graph, so that you can emphasize small differences in values between different categories, make it somehow obvious that you are zooming in. Ideally, only zoom in when the emphasis is on the differences between the data sets (the 'delta'), not so much on the relative ratios.

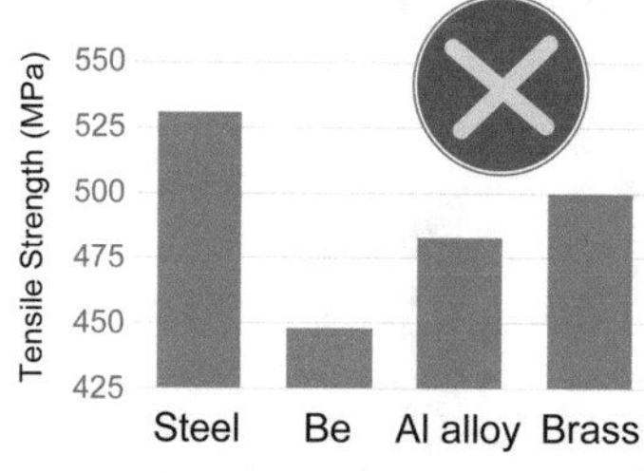

With this choice of vertical axes, it looks like the tensile strength for Brass is three times higher than that for Be, while this is not the case

This more clever choice of vertical axes provides an accurate quick overview of the relative tensile strength values, appropriately demonstrating that while Brass is indeed stronger than Be, the difference is not that large

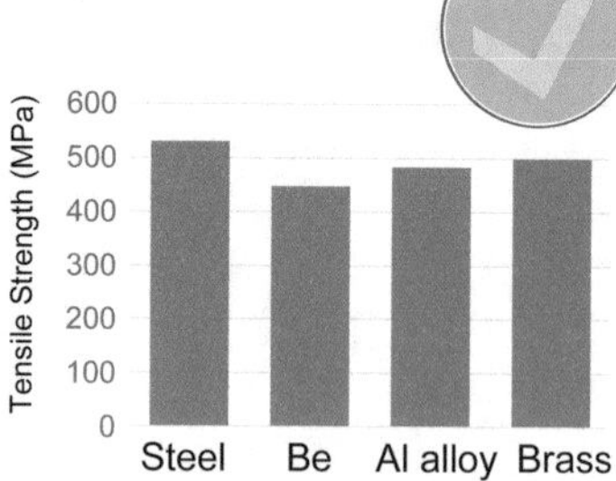

6.10 Animate Your Graphs

Understanding a graph takes time. Even for very smart people familiar with your work. Often times, speakers are deep into explaining critical findings from their graph, while the audience is still trying to figure out the basics (what is being plotted, what's on each axis, the units, the legend). We are thus back with a scenario where the audience is no longer listening to the speaker and rather find him/her annoying. My strong advice is to ANIMATE your graphs:

- First introduce the axes with axes names and units, without any data yet. You can even introduce only one axis at a time
- With the axes defined, introduce only one set of data at a time
- For instance, if you compare results from three different experimental configurations, introduce only one at a time with its corresponding legend
- Spend as much time as needed prior to proceeding with the next data set (animate as slowly as needed)
- Introduce trend lines after you have introduced the data points
- To emphasize specific data points, do not use the laser pointer (Sect. 13.4). Instead, animate some circles or other geometrical figures or pointing arrows to emphasize your data
- Consider making some data disappear after you have discussed it, to avoid cluttering your graph when introducing additional data that does not require the previous one to still be seen

Advantages of animating graphs:

- The audience is not overwhelmed by too much info flashed at once
- The audience is guided through all basic elements of the graphs, in an appropriate sequence determined by the speaker
- Verbal explanations are provided at the same time as visual information is introduced (see Sect. 2.2)

Don't think you are treating your audience like kids when you take the time to slowly introduce your graphs. Quite the contrary, you are demonstrating respect!

Two examples are shown on the next page, where the graphs are introduced in 4 distinct steps each.

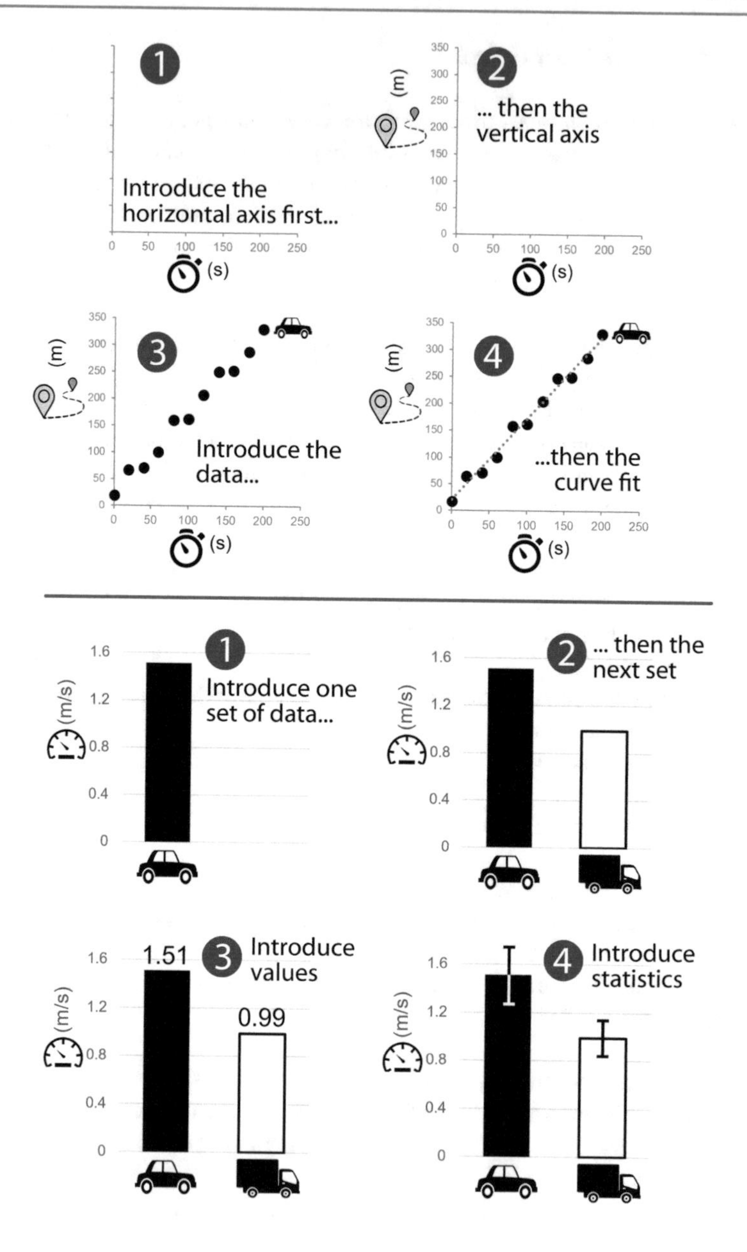
1
Introduce the
horizontal axis first...
0 50 100 150 200 250
(s)
2
... then the
vertical axis
(m)
350 300 250 200 150 100 50 0
3
Introduce the
data...
4
...then the
curve fit
1
Introduce one
set of data...
1.6 1.2 0.8 0.4 0
(m/s)
2
... then the
next set
1.51
3
Introduce
values
0.99
4
Introduce
statistics

6.11 Show Credibility

You might want to share a graph from a famous journal authored by a Nobel prize winner for credibility purposes. Unfortunately, graphs from original sources are generally NOT appropriate for on-screen presentations:

- Fonts might be too small (typical for a scientific paper in print)
- Image quality might be poor (poor scanning from an old article)
- Way too much info shown – you may only care about a few things on it

Here is some advice:

- Show the original graph, mentioning its source (reference, author) to emphasize its trustworthiness
- After a few seconds, grey it out (replace it by an identical greyed out version at the exact same spot – see Sect. 4.10)
- You then superimpose your own graph on top of it, without removing the original greyed out original:
- First introduce axes (on top of the original ones)
- Then highlight only the information you want to convey, superimposed on the original graph
- Then, just remove the original greyed out graph. Only your own variant remains, which is now perfectly suited for your purpose, while still clearly inspired from the famous original

---

Here is an example where a graph is a low-quality scan in a foreign language, with even some hand-written symbols. See on the next page how the information can be made clearer and nicer while still showing the original graph

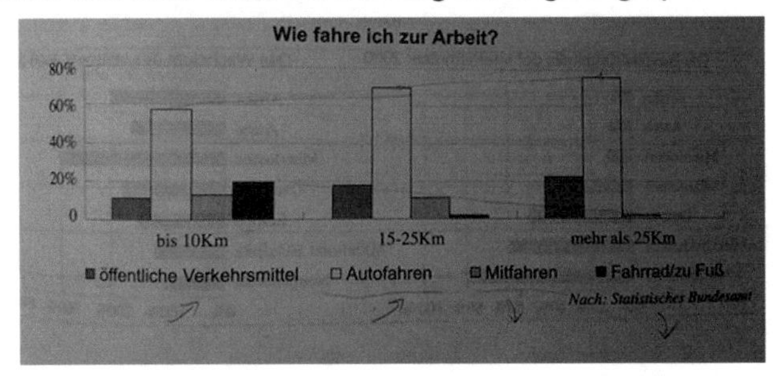

Source: © Statistisches Bundesamt (Destatis), 2020

In a first step, the bars are made clearer, with more vivid tones. And icons (Sect. 6.7) are introduced to replace the legend that was not conveniently laid out in the first place

As the last step, the useless legend for the transportation means is removed, and the horizontal axis values have been moved Finally, the axes have been redrawn, and the original graph has been completely removed

The end result is a much clearer graph meeting most of the guidance from this book, which still offers, through this animation, a tie to the original graph originating from another work/researcher

6.12 Use Colors in Graphs

The graphs presented so far (Sects. 6.1 to 6.11) were all in black and white (greyscale), for a few reasons:

- Color can represent an issue for color-blind people
- Some versions of this book might get published in black and white
- Leaving color aside allowed focusing on other graph features

But it is now time to introduce colors in graphs, as the use of color does definitely improve the visual appearance of graphs, and the ease of understanding and tracking.

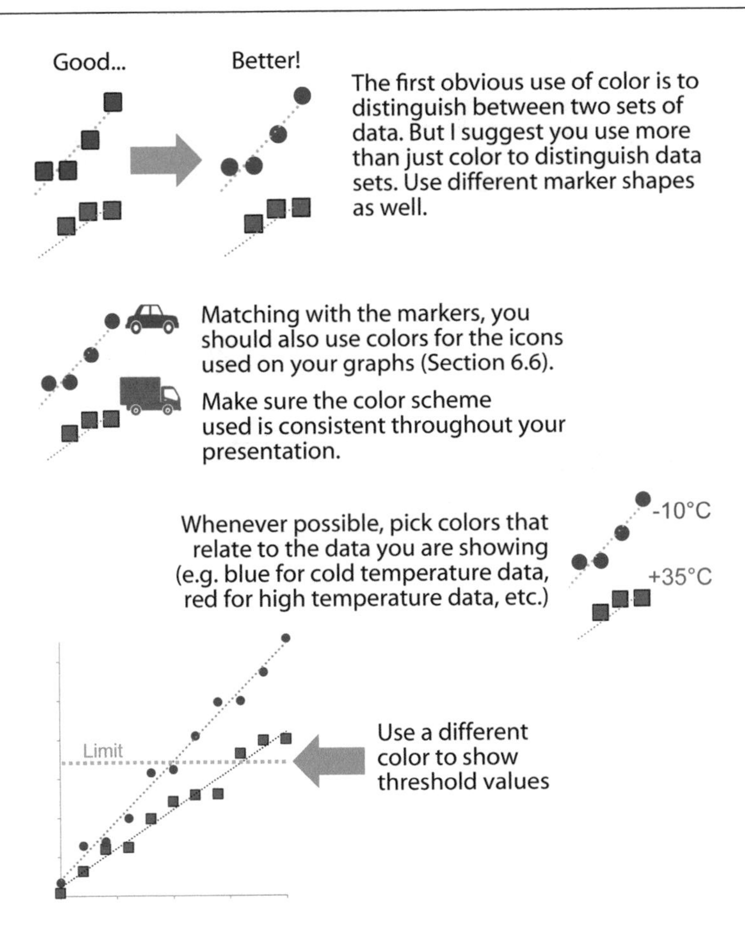

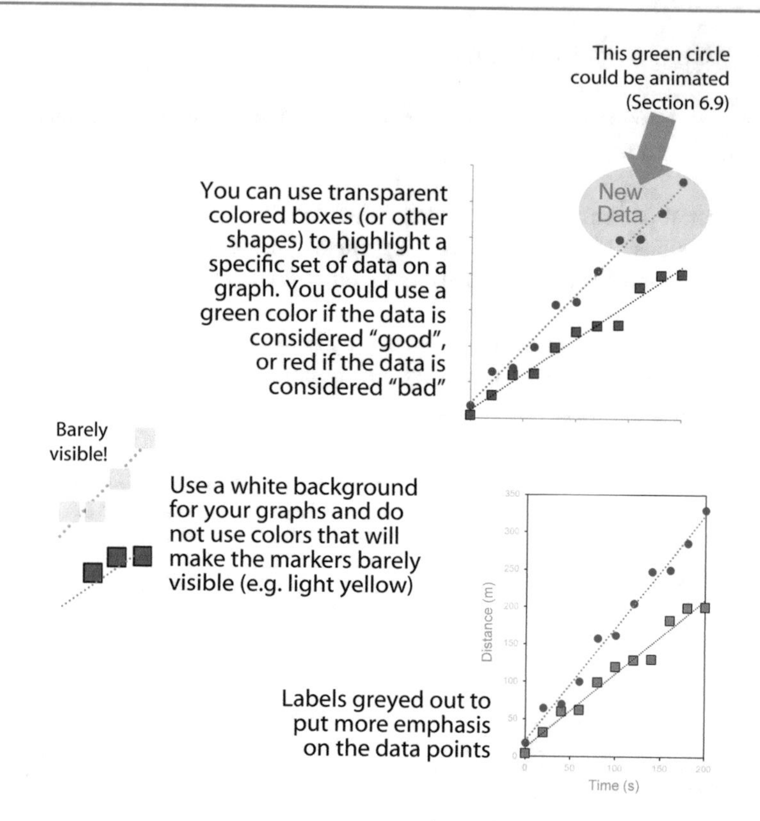

Do not use color just for the sake of using color (Sect. 2.6). No need to use color for the axes or for the labels. On the other hand, as mentioned in Sect. 6.5, you can use grey to downplay the importance of the grids, axes and labels, if the main objective of your graph is to highlight trends as opposed to specific numerical values.

6.13 Change the Graph's Aspect Ratio

Whether a graph is to occupy the entire screen, or just a fraction of it (assuming room is shared with other elements, maybe an image or a bullet list), you should aim at optimizing its size by having it match the room available on the screen.

There is no rule stating that original graphs used in test reports, papers or even in previous presentations should be used as is. If your original graph has a rectangular shape, but the room available on the screen is actually a square, change the aspect ratio of your graph to become a square! Ideally, do this within whatever graph-generating software you normally use.

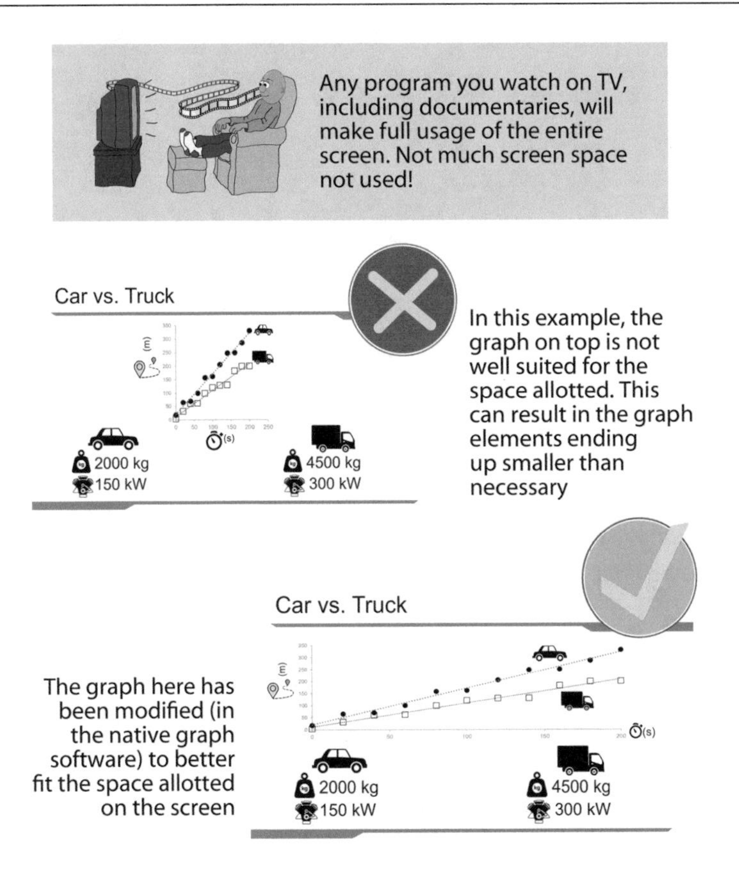

7 Tables

J.-P. Dionne, *Presentation Skills for Scientists and Engineers*,
https://doi.org/10.1007/978-3-030-66069-7_7

7.1 Replace Tables by Graphs

If an image is worth a thousand words, a graph is certainly worth quite a few tables! Whenever you are about to introduce a table in your presentation, ask yourself whether that table could be represented in a graphical format instead. If you can indeed find a graphical way to present your data, go for it. And don't be shy to use animations! (See Sect. 7.3).

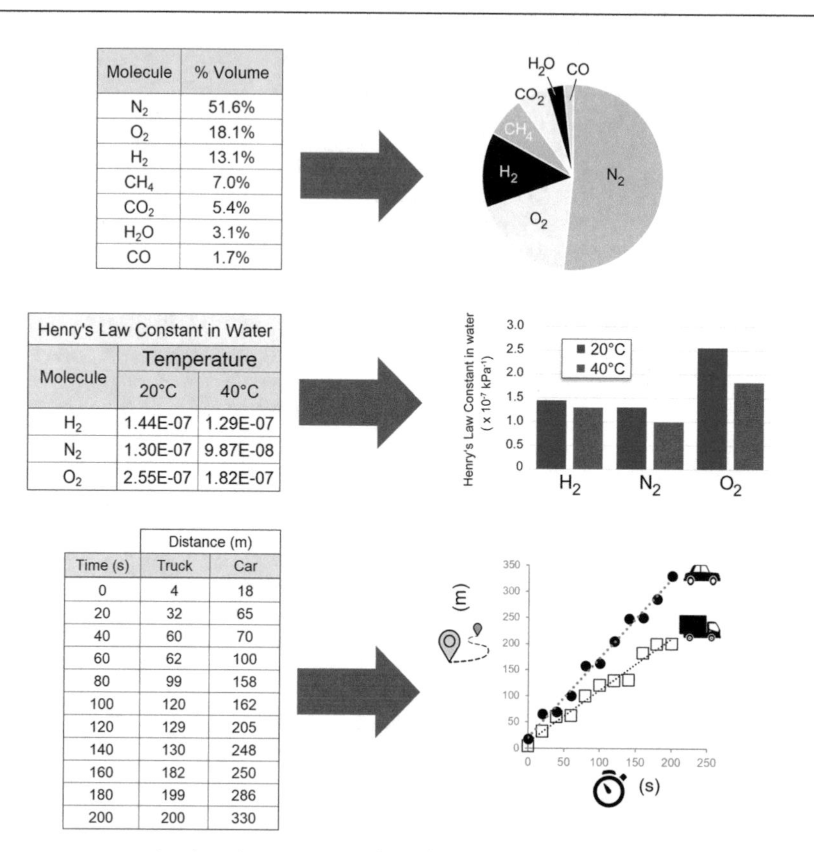

Molecule	% Volume
N_2	51.6%
O_2	18.1%
H_2	13.1%
CH_4	7.0%
CO_2	5.4%
H_2O	3.1%
CO	1.7%

Henry's Law Constant in Water		
Molecule	Temperature	
	20°C	40°C
H_2	1.44E-07	1.29E-07
N_2	1.30E-07	9.87E-08
O_2	2.55E-07	1.82E-07

	Distance (m)	
Time (s)	Truck	Car
0	4	18
20	32	65
40	60	70
60	62	100
80	99	158
100	120	162
120	129	205
140	130	248
160	182	250
180	199	286
200	200	330

In the three examples shown, get rid of the tables, and use the graphs instead

7.2 Create Tables "Externally"

The PowerPoint built-in table creation tools lack the flexibility and advanced features from Excel. Moreover, if you already have Excel tables, you might not feel like re-building them. As such, feel free to import Excel tables in your presentations.

Three ways to "import" Excel tables in PowerPoint:

1. "Paste" – Select and copy tables in Excel, and then paste them within PowerPoint
   - Disadvantage: The look and format is likely to be altered
   - Advantage: Edits can be made within PowerPoint
2. "Paste Special" (Windows/Enhanced metafile) – The format of your tables will not be altered (tables imported as images)
3. Screen Capture (see Sect. 4.5) – Table format unchanged:
   - Prior to hitting "Print Screen", remove gridlines within the page
   - For best image resolution, increase table size to fill the screen
   - Once pasted within PowerPoint, "crop" the table to remove unwanted areas around it

Disadvantages of the last two options:

- Any edits must be made within Excel, tables must then be re- imported into PowerPoint (no big deal if content/format is final)
- A white area around the table might be left if cropping is not made properly (not an issue with a white template background)

Material	Yield strength (MPa)	Tensile strength (MPa)	Density (g/cm³)
Steel	247	841	7.58
Acrylic	72	87	1.16
HDPE	26–33	37	0.85
Polypropylene	12–43	19.7–80	0.91
Cast iron	130	200	7.3
Beryllium	345	448	1.84
Aluminium alloy	414	483	2.8
Copper	70	220	8.92
Brass	200 +	500	8.73
Tungsten	941	1510	19.25
Aramid	3620	3757	1.44

Material	Yield strength (MPa)	Tensile strength (MPa)	Density (g/cm³)
Steel	247	841	7.58
Acrylic	72	87	1.16
HDPE	26–33	37	0.85
Polypropylene	12–43	19.7–80	0.91
Cast iron	130	200	7.3
Beryllium	345	448	1.84
Aluminium alloy	414	483	2.8
Copper	70	220	8.92
Brass	200 +	500	8.73
Tungsten	941	1510	19.25
Aramid	3620	3757	1.44

Table can be edited, but the look and format is very different

Material	Yield strength (MPa)	Tensile strength (MPa)	Density (g/cm^3)
Steel	247	841	7.58
Acrylic	72	87	1.16
HDPE	26–33	37	0.85
Polypropylene	12–43	19.7–80	0.91
Cast iron	130	200	7.3
Beryllium	345	448	1.84
Aluminium alloy	414	483	2.8
Copper	70	220	8.92
Brass	200 +	500	8.73
Tungsten	941	1510	19.25
Aramid	3620	3757	1.44

Original worksheet

Option 2
"Paste Special"

Material	Yield strength (MPa)	Tensile strength (MPa)	Density (g/cm^3)
Steel	247	841	7.58
Acrylic	72	87	1.16
HDPE	26–33	37	0.85
Polypropylene	12–43	19.7–80	0.91
Cast iron	130	200	7.3
Beryllium	345	448	1.84
Aluminium alloy	414	483	2.8
Copper	70	220	8.92
Brass	200 +	500	8.73
Tungsten	941	1510	19.25
Aramid	3620	3757	1.44

The result is generally quite good,
but the table cannot be edited

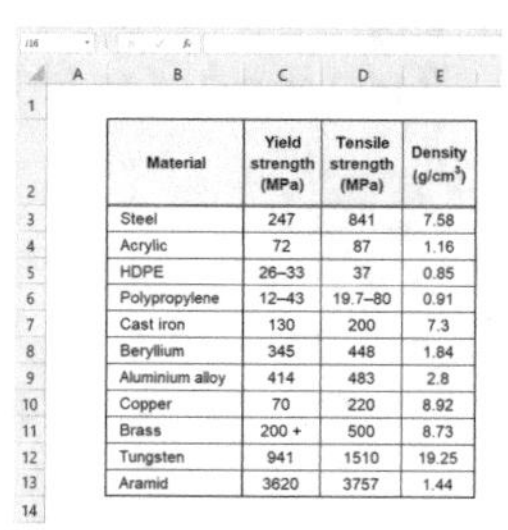

Material	Yield strength (MPa)	Tensile strength (MPa)	Density (g/cm^3)
Steel	247	841	7.58
Acrylic	72	87	1.16
HDPE	26–33	37	0.85
Polypropylene	12–43	19.7–80	0.91
Cast iron	130	200	7.3
Beryllium	345	448	1.84
Aluminium alloy	414	483	2.8
Copper	70	220	8.92
Brass	200 +	500	8.73
Tungsten	941	1510	19.25
Aramid	3620	3757	1.44

Original worksheet

Option 3
Screen
capture
and crop

Material	Yield strength (MPa)	Tensile strength (MPa)	Density (g/cm^3)
Steel	247	841	7.58
Acrylic	72	87	1.16
HDPE	26–33	37	0.85
Polypropylene	12–43	19.7–80	0.91
Cast iron	130	200	7.3
Beryllium	345	448	1.84
Aluminium alloy	414	483	2.8
Copper	70	220	8.92
Brass	200 +	500	8.73
Tungsten	941	1510	19.25
Aramid	3620	3757	1.44

Guarantee: table looks exactly
as per the original
Image size always reasonable
But the table cannot be edited

7.3 Animate Your Tables

Similar to graphs (Chap. 6), tables may contain a lot of information, requiring time for the audience to grasp. If a complex table is not introduced gradually, the audience will try understanding the table while losing track of what the speaker says.

My strong advice is to ANIMATE tables:

- First introduce column names (with units, when relevant), and then the row names (when relevant). Do not introduce the content yet.
- With column and row names defined, introduce only one set of data (row or column) at a time:
  - For instance, if you compare results from three different experimental configurations, introduce only one configuration at a time
  - Spend as much time as needed prior to proceeding with the next data set
- To emphasize specific table elements, do not use the laser pointer (Sect. 13.4). Instead, animate some rectangles or other geometrical figures or pointing arrows to emphasize your data

Advantages of animating tables:

- The audience is not overwhelmed by too much info flashed at once
- Verbal explanations are provided at the same time as visual information is introduced (Sect. 2.2)

When animating tables created within a spreadsheet software, multiple variants of the table might need to be generated, with increasing content. Alternatively, you can hide (for instance with white opaque rectangles) elements from your table, and then make these rectangles disappear at the appropriate time, using animation tools.

Similar to graphs, don't think you are treating your audience like kids when you take the time to slowly introduce your table. Quite the contrary, you are demonstrating respect.

Step 1: Introduce column names. Take your time to describe them

	Theory	Numerical Simulations	Experiments

Step 2: Introduce row names. Take your time to describe them

	Theory	Numerical Simulations	Experiments
Config 1			
Config 2			

Step 3: Show the first set of data, and comment as needed prior to showing the rest

	Theory	Numerical Simulations	Experiments
Config 1	3.50	3.51	3.48
Config 2			

Step 4: When done describing the first set of data, show the next set. Now is the time to compare the two sets

	Theory	Numerical Simulations	Experiments
Config 1	3.50	3.51	3.48
Config 2	4.25	2.26	3.99

Step 5: To assist with the discussion, feel free to highlight values in the tables that deserve further discussion

	Theory	Numerical Simulations	Experiments
Config 1	3.50	3.51	3.48
Config 2	4.25	2.26	3.99

Do not hesitate to use colors in your tables, mostly to highlight specific values (font color, or background color). But just like for graphs and anything else, do not use color just for the sake of using color!

7.4 Use Icons in Your Tables

A basic table only contains text and possibly numbers. But to assist the audience with a better and quicker understanding of your table, similar to what has been recommended for graphs (see Sect. 6.7), do not hesitate to use icons and images, which can also be animated.

Examples:

- To represent "time", you can show a clock icon
- To represent "distance", you can use a measuring tape icon
- To represent "temperature", show a thermometer next to the data
- If one set of data corresponds to a "car", show a car icon

Use very simple images as icons, ideally black and white silhouettes (see Sect. 4.4).

Consider replacing all text with images and icons, so long as the meaning remains obvious (this also applies to legends).

	Spring	Summer	Fall	Winter
Min Temperature (°C)	12	24	8	5
Max Temperature (°C)	29	40	24	22
Rain / Snow (mm)	8.4	29.0	19.6	26.2

°C	12	24	8	5
°C	29	40	24	22
mm	8.4	29.0	19.6	26.2

8 Maths

J.-P. Dionne, *Presentation Skills for Scientists and Engineers*,
https://doi.org/10.1007/978-3-030-66069-7_8

8.1 Avoid Maths and Equations

Similar to graphs, mathematical equations can take a very long time to appropriately introduce. They also demand a significant intellectual effort. As such, equations are likely to be a turn-off for the audience.

A few tips:

- Only show equations critical to your objectives and conclusions
- Use a large enough font
- Never introduce more than one equation at a time
- Ensure that every single term is well defined
- Use animations to highlight each term step by step (use arrows, one-word descriptions, etc.)
- Avoid deriving equations during a talk. The audience will trust that they can find such details in your paper, or discuss with you off-line
- Don't feel the need to prove/derive everything on the screen

If you really need to show a whole bunch of equations to demonstrate the credibility of your approach, I suggest you just flash the equations, emphasizing a few key points maybe, apologizing for having to show them. This is one rare case where the audience will most likely not be mad at you for not discussing in detail something shown on the screen!

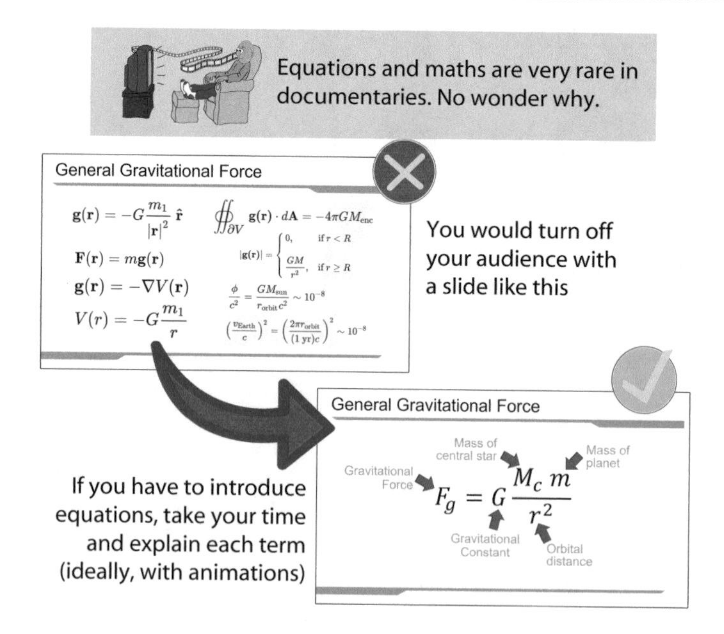

8.2 Use Just Enough Significant Digits

Do you really need to provide numbers accurate to 0.0001? Only use relevant significant digits!

- Graphs created in spreadsheet software sometimes contain useless digits on the axes (e.g. 1.00, 2.00, 3.00 instead of 1, 2, 3). Make sure you set that up properly
- Equations from trend lines created in spreadsheet software also often contain way too many significant digits. Fix that directly on your graph, or within your slide (overwrite the info)
- In tables, use the minimum number of significant digits that will suffice to clearly differentiate all values. Especially given that using too many digits will result in having to use smaller fonts to fit everything

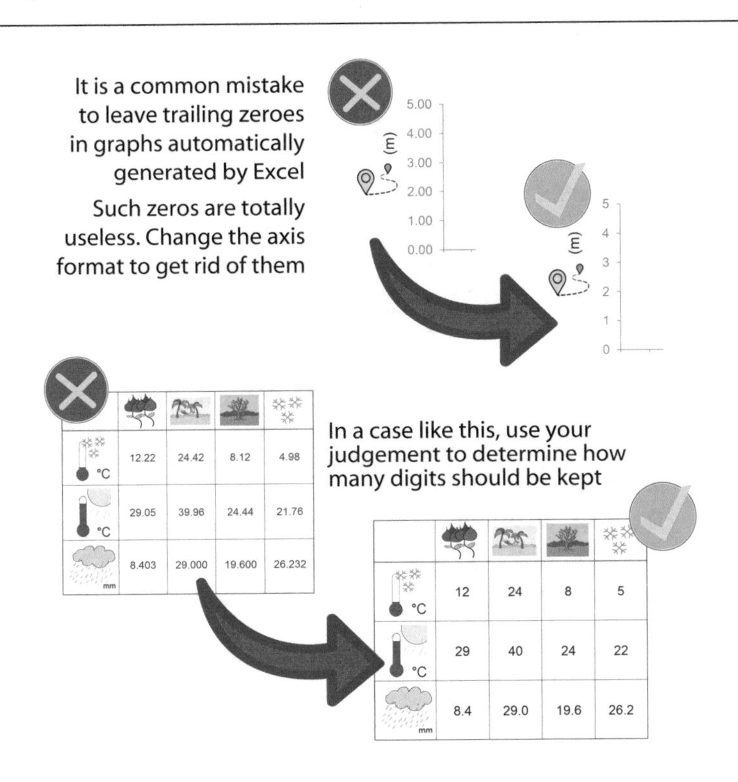

8.3 Put Values in Perspective

In a presentation, let's say you mention that Business R&D expenditures in Canada amount to as much as 16.9 billion dollars. Wow, that seems large! But compared to what? Beyond some level, numbers become meaningless. When dealing with such large numbers, and even when using smaller ones, use percentages, side by side comparisons, etc. All the values you insert should be put in perspective. Don't leave numbers on their own!

Don't just let the audience "feel" a number is large. Prove it, by putting it in perspective! By the way, such "mistakes" are also found regularly in prestigious publications.

This is equivalent to the "rule" of including a coin or a ruler in a photograph, for the audience to appreciate the relative size of the image you want to show.

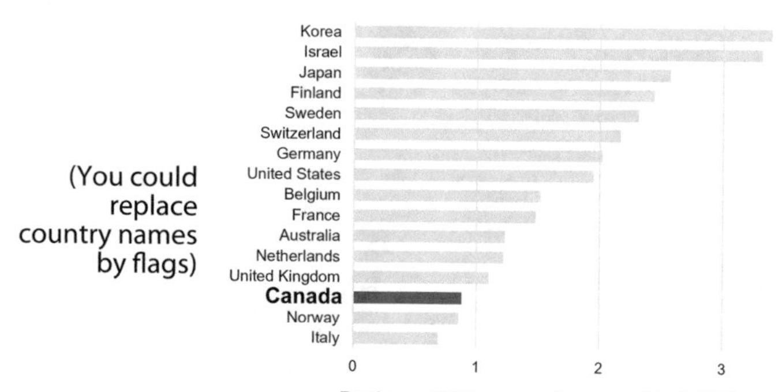

Rather than just stating that Business R&D expenditures in Canada amount to 16.9 billion dollars per year, put that value in perspective.

The graph above does it in a brilliant way, by comparing with the GDP (percentage value), and also by comparing with what is done in other countries. All of a sudden, that 16.9 billion dollar value no longer appears to be that large!

8.4 Really Need Equations?

As mentioned in Sect. 8.1, you should avoid equations as much as possible in your slide deck. But if mathematical development is central and critical to your work, take advantage of existing software tools to "type" equations directly on the computer.

If your slide presentation software (PowerPoint or other) has a built-in equation editor, explore its features. For instance, the built-in PowerPoint editor allows you to write equations directly using your mouse, which get "translated" to numbers, variables and symbols in real-time (see image below). That's especially convenient if you do not feel comfortable using the various menus and options.

This being said, also explore other mathematical packages available on the market, as they all have strengths and specialties, either for ease of editing or aesthetics. Each software tool allows exporting created equations to slide presentation software. In addition, keep in mind that with any software package, the "screen capture" option is always available (Sect. 4.5).

Some of these tools are relatively general in nature, allowing both equations and other types of texts to be generated (e.g. Markdown, LaTEX, LyX), while other specialize in mathematical objects (e.g. EquatIO, MathMagic, MathCast, and MathJax). Detailed explanations of how these tools work would go beyond the scope of this book, but see short descriptions on the next page.

If you already use one of these tools, there is no point in generating new equations from scratch within your slide presentation software.

Finally, nothing stops you to simply use a pen and paper to write down your equations with whatever colors, extra marks, etc. and take a picture of it with your smartphone. The final result might not look as "professional", but it adds a personal touch, plus lots of flexibility!

The PowerPoint built-in equation editor allows you to use your mouse to directly type in equations, which automatically get "translated" in realtime

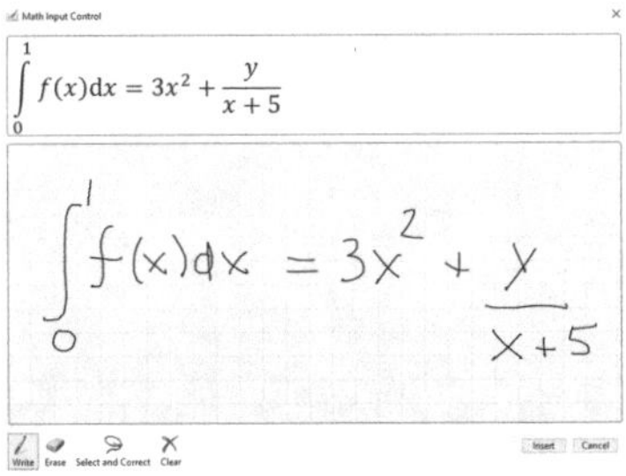

LaTEX, Markdown and LyX are powerful software packages allowing for various document contents to be created, including mathematical equations

EquatIO focuses on education

MathMagic (which has a Pro version) features various keyboard shortcuts, and batch tools

MathCast
-the open source equation editor

MathCast is a free and open-source equation editor

MathJax focuses on generating equations for web browsers. Images can readily be scaled without loss of resolution

8.5 Manage Metric vs. Imperial Units

In most scientific circles, the metric system is the only one used. However, when presenting outside of purely scientific circles to an audience which is American or mostly American, imperial units should be considered. Find out what is expected from any US audience to determine whether to show imperial units, metric units, or a combination of both.

When showing numbers with both sets of units, I suggest you show the metric ones first, and then the imperial ones in brackets, in a lighter shade of grey.

When transferring graphs from metric to imperial units, do not just change values on the axes to reflect the revised units. For instance, do not just change axes from [1 m, 2 m, 3 m] to [3.28 ft., 6.56 ft., 9.84 ft]. Instead, have the gridlines line up with feet values [1 ft., 2 ft.… 10 ft]. To achieve this, raw values must be modified in the associated tables.

If you are to show both sets of units, I suggest building your graphs in metric units, and then add imperial unit values in a lighter shade of grey on the top and right of the axes.

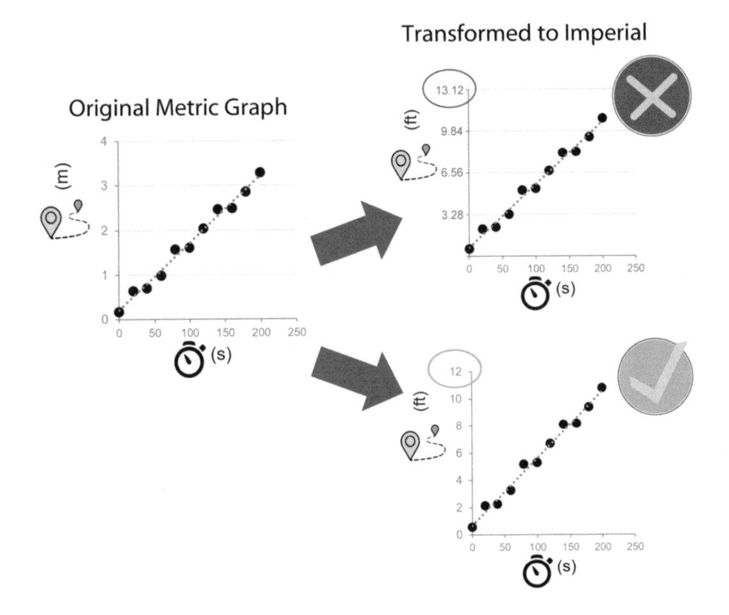

9 Structure

J.-P. Dionne, *Presentation Skills for Scientists and Engineers*,
https://doi.org/10.1007/978-3-030-66069-7_9

9.1 What to Do with the Cover Page?

Do you really need to read your title out loud? Do you really need to introduce yourself?

No need to thank the conference organizers. Imagine all speakers doing the same. Will start sounding like an Oscar ceremony, in a more boring variant.

Just dive into it!

You flash the cover page, and then move on to the next slide. Dive into it.

Or… just kick-off your talk with the cover page as the background. That's actually a good way to connect with your audience, since they will really focus on you at that stage, as opposed to your slide, which they have seen and read already.

Documentaries, like other TV programs, have a nice intro with nice graphics and music. But is there a narrator reading out loud the name of the show? And very often, a documentary starts with a closeup of the presenter, without any visual aid. This helps creating a connection. You could do the same with your presentation.

9.2 Ditch the Table of Contents

Chances are, you will start with an introduction and then proceed with some background, past research, results, and end with conclusions. You will then open the floor to questions. Pretty much in that exact sequence. Do you need a slide and can you afford wasting precious seconds to state the obvious?

Your audience might benefit from a table of contents to have the presentation's structure in mind. But there is a better way.

Let's say you present experimental results, which highlighted problems. To address these problems, you conducted numerical simulations. Finally, you established correlations between the experimental and numerical work, and you drew some conclusions. You could have the following table of contents displayed:

- Background
- Experimental Test setup
- Results
- Numerical simulations
- Discussion and Conclusion

But much more effective, and much more entertaining, would be to introduce, one at a time:

- A representative image of your experiment or experimental setup
- A representative image of your numerical simulations
- A two-way arrow pointing between the two

No text. And the audience knows what you plan on talking about.

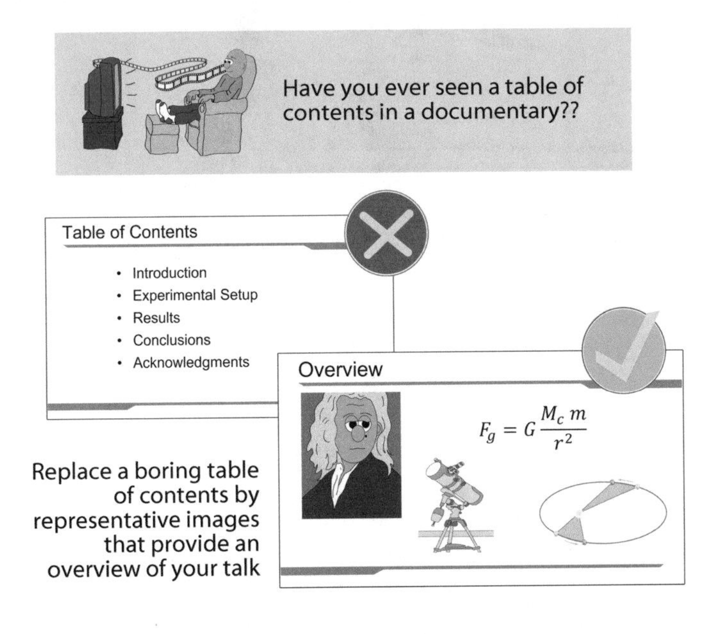

9.3 Keep Track of Progress?

Take the example of that book you currently hold in your hands (or read on the screen): it includes section names on the header of each page (e.g. Introduction, Animations, Graphs, Structure, etc.) Such a feature is useful in a reference book, to assist readers finding the information they need, when looking for something specific.

But do you need such visual reminders of which 'section' you are up to, in a presentation? For instance, should you indicate on every slide, whether you are in the 'Introduction' or 'Results' section?

One might feel that the audience might appreciate being able to 'gauge the progress'.

However, there is an important difference between a reference book and a presentation. Granted, in a book, one has a good feel for how much is left to read just by looking at the book as a whole. But the main difference is that in a presentation, the audience is guided and taken along. The audience does not need to search through your presentation to find a specific section. Plus, unless you are doing a poor job as a presenter, it should be obvious whether you are presenting background info, results or a conclusion.

Including sections in a presentation might be useful for long lectures, especially if the slides eventually become reference materials that get shared with students. But for a time-limited presentation (as in conferences), there really is no need to add that type of information, which would eat away at the limited room available on your slide.

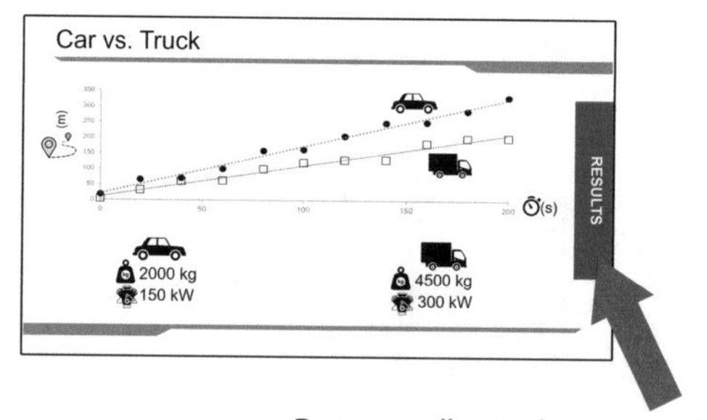

Do you really need to emphasize that you are in the "Results" section? Isn't it obvious?

9.4 Use Appropriate Slide Titles

Space is often limited on slides, especially given that text and images must be large enough to be appropriately seen by the audience. As such, slide titles should be carefully drafted to convey critical information directly linked to each individual slide. You should not keep the same slide title for a few slides in a row (e.g. 'Results'). That would only help keeping track of the section of the presentation you are currently in (See Sect. 9.3), which is not that helpful.

Take this book for instance: almost every page has a different title, directly relevant to the associated text, giving a clear indication of what to expect on each page. How much would be gained by replacing these dedicated titles by the more generic 'section titles', such as 'Introduction'? Plus, if you keep indicating 'section titles' such as 'Introduction', you risk ending up having to add a sub-title to each slide, losing some precious space. Just drop the 'section title' and just focus on that 'sub-title'!

Let each slide title be unique and specific! Just make sure the titles remain short (do not spill over two lines). You can even consider not including slide titles, at least for some slides. They might not always be needed.

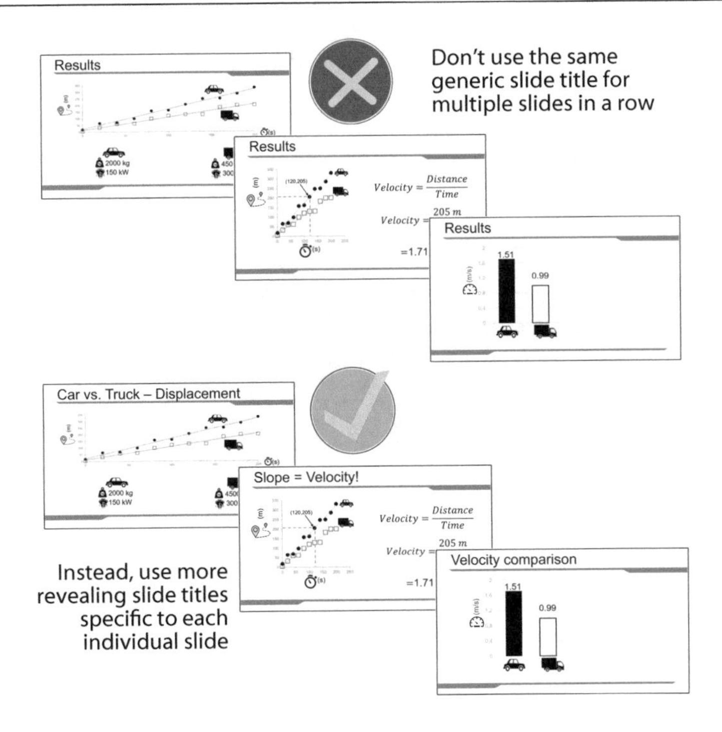

9.5 Start with a Question, or Odd Fact

A key challenge as a presenter is to engage the audience right from the start. You have to let them know exactly what your topic is and where you are heading (Sect. 9.6). But beyond that, you should also manage to pique their interest.

You can start with a question for the audience, not to be answered "vocally" and right away, but rather to have it "contaminate" their mind. You then provide the answer by the end of your talk. If the question is interesting and challenging enough, the suspense generated will keep the audience riveted to the screen!

An interesting variant is to turn that question into a "quiz" with multiple-choice questions. Once again, answers are to be provided only towards the end of your talk. But of course, as you present your material, the list of clues will grow (if your presentation is well- built). Everybody likes quizzes, and we all like to do well!

An alternative to quizzes is simply to bring up some interesting or counter-intuitive facts related to your presentation topic ("Did you know that (…)?"). That should work too!

9.6 Have Clear Objectives and Scope

The audience should not still be wondering what exactly you are aiming at, in the midst of your talk. Too many speakers dive deep way too early in their detailed results and theory, prior to having clearly introduced the objectives and scope of the study.

For the benefit of the audience (as well as your own), clearly introduce the questions you want to address in your talk. Ideally, that should be just a few key points. Otherwise, I suggest you down select fewer topics and thus reduce your presentation scope.

The few key points you introduce early on will be the same key points that will be addressed again in your conclusion slides. There should be a perfect match between the objectives listed at the onset of your talk and the conclusion points.

The one question you do not want people in the crowd to ask themselves is: "what is the point of all this"? While your results and theory might potentially be quite complex (which makes sense if you are presenting to subject matter experts in a highly scientific forum), make the objective and scope sections extremely easy to understand, even for those not familiar at all with your world.

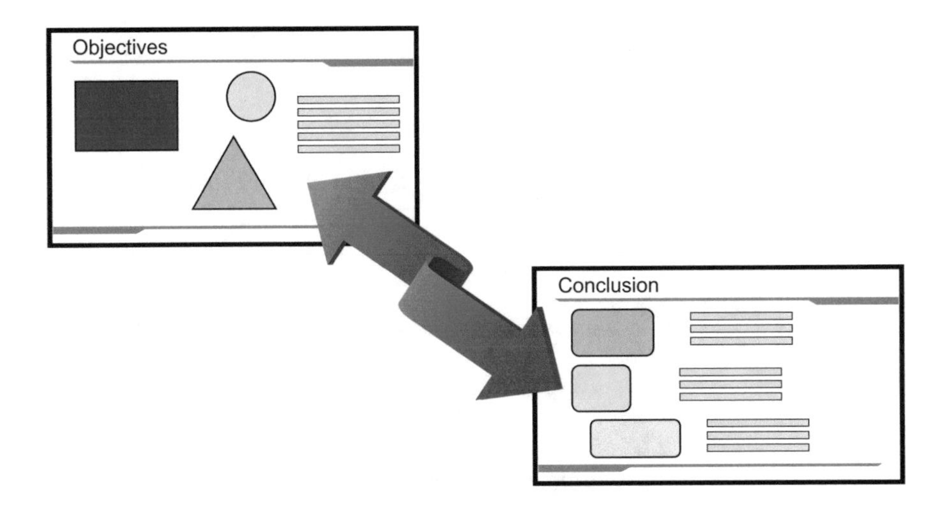

9.7 Use "Flashbacks" as Conclusion

Your conclusion slides typically do not introduce new information and concepts, but rather summarize the key findings presented. A quick and effective way to summarize and re-emphasize the key concepts from your talk, is to introduce a key image for each main finding or conclusion. Two interesting options suggested:

Option 1: reproduce a smaller version of a previous graph or image from your talk, which will immediately remind the audience of how the key concept was originally introduced. The graph should look the same, but since it will be a smaller version, you could remove some details (e.g. numbers of the axes) or enlarge text (like axes text)

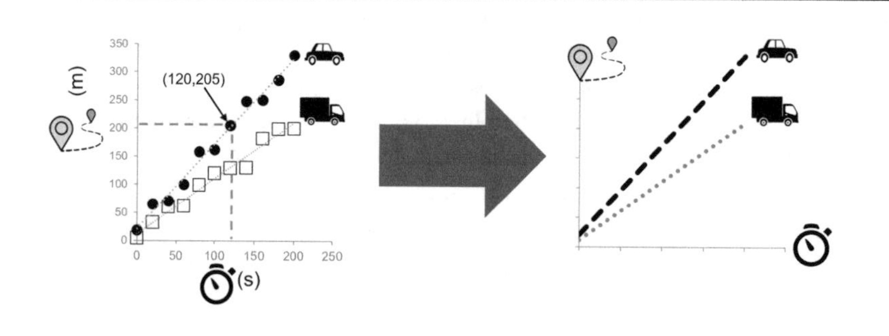

Option 2: Even simpler: display small screen captures of entire slides to illustrate each concept

Summary

Displacements were measured

Velocities calculated (slope)

Car is ~ 50% faster than truck

9.8 Use Last Slide for Acknowledgments

Rather than giving credits to your co-authors at the onset of your talk, keep that to the last slide. On your parting slide, write down your acknowledgment statements as text (this and citations are the only occasions where I suggest you actually include more than a few words of text). This way, even if you are short with time, your acknowledgments will be conveyed. By having your acknowledgments written on the last slide, you know that it will be shown on screen for more than just a few seconds. Even consider including photos of your co-authors or colleagues you wish to acknowledge (with their permission).

An advantage of introducing your co-authors at the end of your talk, is that it might deflect complicated questions towards them, if they happen to be in the audience!

Alternatively, you can use your last slide to leave a few parting or take away points, which the audience can read while they applaud or during the question period.

10 Interactions

J.-P. Dionne, *Presentation Skills for Scientists and Engineers*,
https://doi.org/10.1007/978-3-030-66069-7_10

10.1 Get Interactive! Use Action Buttons

In most presentations, speakers only perform one of two "actions": moving up or down a slide (or animation) using the keyboard (arrows, space bar), or through a remote tool.

But many more "live actions" can be performed. Consider the use of "Action Buttons", which can be activated by clicking or hovering (passing over with the mouse) during your talk, to:

- Directly move to a specific slide (rather than flipping through multiple slides and animations)
- Move back to the previously shown slide (where you started)
- Open a weblink
- Run a program/open a file

Such actions can make your presentation more interactive and provide flexibility rather than having to stick to a rigid scenario.

Even for short presentations without audience interaction, action buttons can still be used to:

- Skip slides if you realize you are running out of time
  - It is frustrating for the audience to see slides being flashed for only a fraction of a second – show a slide well, or don't show it!
- Show additional slides if you have more time than anticipated

Slide presentation software provides a number of built-in Action Buttons.

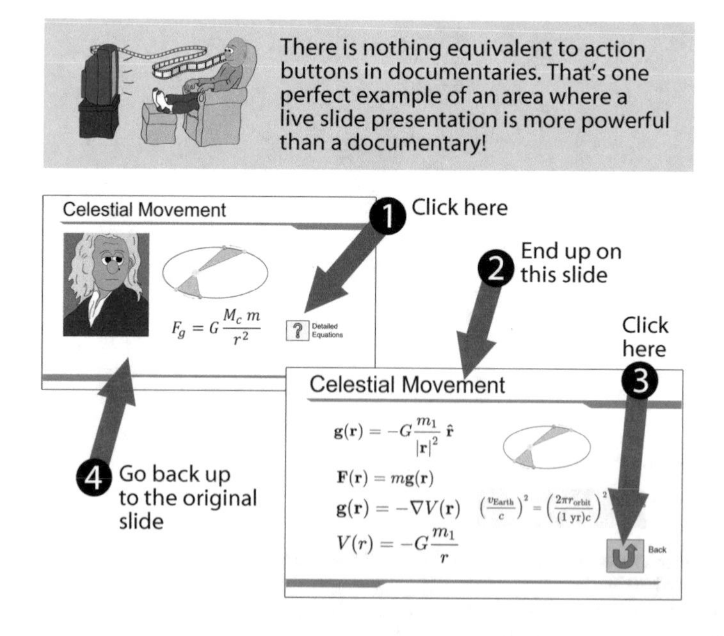

10.2 Use Existing Slide Elements as Action Buttons

Slide presentation software provides built-in action buttons (Sect. 10.1). However, if you show an action button which you end up not using, the frustrated audience will wonder what that button was.

Instead, consider using graphics or text already included in your presentation as action buttons. Such existing elements do not look like action buttons and as such, they will not attract the attention from the audience. If you do not use them, no one will notice.

Another advantage of using existing elements as action buttons is a potentially closer link with the actions associated with specific elements. For instance, a graph can be used as a button to provide details about that specific graph (maybe as a pop-up, Sect. 3.8). Or a text box can be used as an action button to move to another slide providing more in-depth explanations. Or clicking on a table might open a spreadsheet file with the original data.

To associate actions with existing elements, select an element, and then choose "action" from the PowerPoint menu.

Be careful though, when you click on action buttons during your talk: if by mistake you click at the wrong place, you might end up moving up one slide.

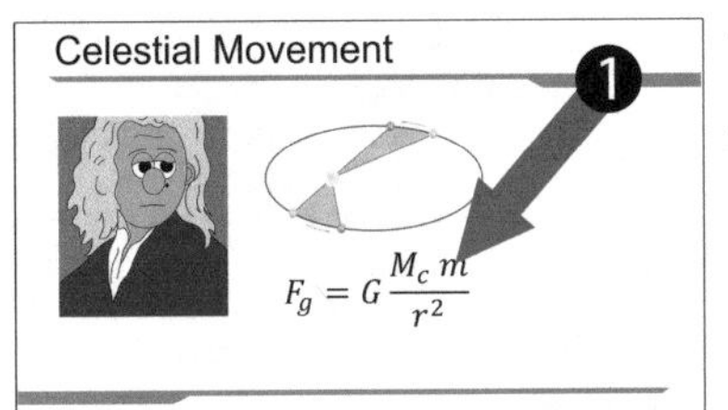

You can use the equation itself as an action button to display the other slide (with the detailed equations)

You can then use the graph as an action button to go back to the original slide

Same example as in Sect. 10.2, except that elements from the slides are used as action buttons, instead of creating buttons dedicated for actions

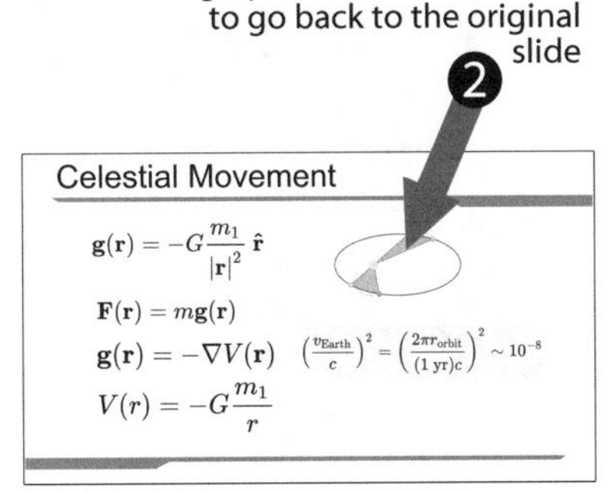

10.3 Take Advantage of Action Buttons During the Question Period

A very convenient use for Action Buttons is to directly move anywhere in your deck, without having to flip through all intermediary slides. Flipping through multiple slides happens the most during the question period at the end of a talk. One person from the audience might demand that you display again a specific graph or photo from the early part of your presentation. To go back there, you could flip through all intermediate slides, or leave the presentation view and search the appropriate slide within your PowerPoint edit screen.

However, a much better option would be to anticipate questions you might get from the audience and introduce action buttons that will quickly take you back to specific slides of interest. Even if you can't anticipate exact questions, having "links" to specific key slides makes it much quicker to navigate back and arrive at destination.

The best way to introduce such "question period" action buttons is to "embed" them in your conclusion slides. Consider using already existing elements (Sect. 10.2) as action buttons. Even better, consider using the "flashback" elements from your conclusion (Sect. 9.7) to associate them with actions.

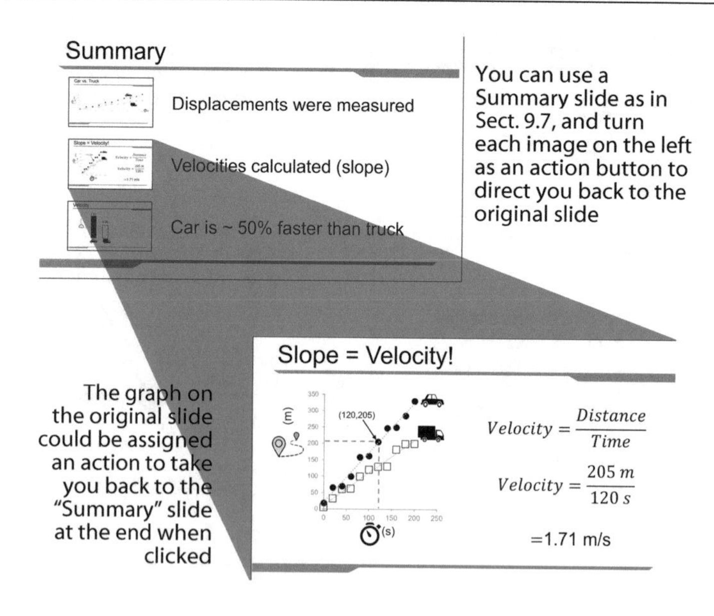

11 Practice

J.-P. Dionne, *Presentation Skills for Scientists and Engineers*,
https://doi.org/10.1007/978-3-030-66069-7_11

11.1 Never Learn Your Text by Heart

Delivering a good presentation, especially a short one with a tight and enforced time limit, requires a lot of practice. Practicing a lot does not mean learning your text by heart. This is actually a very bad thing to do. Delivering verbatim a speech learned by heart always sounds highly non-natural. Plus, if you forget a word or a sentence, you will lose balance and control, and might never get back on track.

However, with a relatively short talk (less than 15 minutes), you may write down the text of your speech, just to clearly determine what you intend to talk about – that helps gauging the time required to deliver the information:

- Print your presentation slides, 6 slides per page. Use this to hand-write your text (very limited space, that's OK)
- As you read and time yourself, you can cut/add material and cleanup sentences, until you can meet the tight timeline
- Writing down and improving the text minimizes chances of dwelling over any particular subject, getting caught in your words
- But never read your text during your talk!

HOWEVER…

- Once you are happy with your text, start practicing WITHOUT IT!
- Go through your slides and give the talk to yourself. At first, you can do it silently, in your head. Ideally, towards the end, you give it out loud
- You repeat the process over and over. In between rounds of practice, you can go back to your text and read sections you had trouble with
- Again, do not try learning your text by heart, even just sections of it. It will show
- As you practice, time yourself, making sure you respect the time limits (minimum and maximum). Analyze sections where you think you are dwelling too long or keep forgetting to mention key elements. Rework your written text as needed

Writing down your text is just a suggestion. If you feel relatively comfortable about your topic, just start practicing right away. But always time yourself.

As you keep practicing, you will realize that quickly enough, you will remember what you want to say for all your slides. When you reach that stage, it is even possible to practice your talk without having your presentation in front of you. That will allow you to be able to practice in your shower, on the bus, while driving, etc.

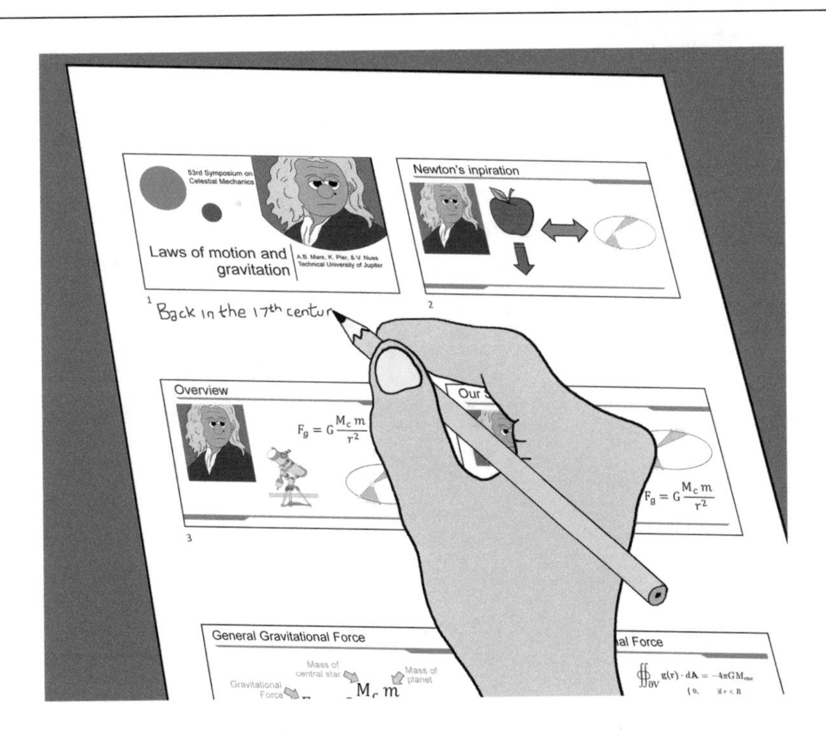

11.2 Short Talk, Long Practice (and Vice Versa)

It might seem like a paradox, but shorter time-limited talks demand more practice than long lectures.

The challenge with short talks is not to end up dwelling on a topic, but rather provide concise explanations to push your points. In short talks with time limitation, every second counts! Dwelling on something will stop you from spending quality time on the remaining slides. And your stress level will jump up the roof as you get the "one minute left" warning with so many slides left to cover!

On the other hand, when delivering a long lecture, you have the opportunity to adjust the level and pace of information provided to match the time allotted. Losing a minute dwelling on something will not have the same severe consequences when giving a long lecture, compared to a quick 5-minute talk.

Plus, it is not realistic to repeatedly practice an hour-long presentation! For long presentations, rather than practicing multiple times, quickly glance through all slides, ensuring that you know precisely what you want to address on each of them.

I still suggest you focus on repeatedly practicing the first slide (see Sect. 11.5), even for long lectures. Once you get started on the right foot, everything flows.

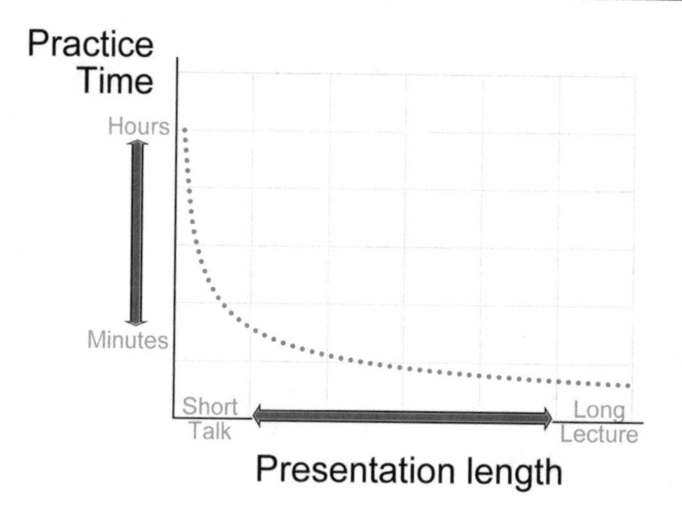

11.3 Second Language? Avoid the Tough Words!

If you have a strong foreign accent, chances are the audience will be forgiving. No need to apologize ahead of time or provide excuses. Just speak slowly and do the best you can. You are not the only one on that boat!

But while improving your general pronunciation skills in a just a few days or weeks is unlikely, you can at least potentially identify a few English terms that give you a lot of trouble. I suggest two options:

- Practice the pronunciation of just these few words. Get some help from a native speaker if you can
- But possibly even better, look for synonyms or alternate expressions

The second option might not always be possible. For instance, if you present on "nuclear physics", chances are you will have no choice but pronouncing the word "nuclear" once in a while (even some native speakers have a hard time with this one!)

If you can find alternative words or expressions, you might not be able to replace all instances of the difficult words (you can leave them as written text), but at least, this will enhance the flow of your talk and you will feel more comfortable.

11.4 Do Not Leave "Untranslated" Stuff!

If your original tables or graphs are in a language other than the language you will use for your presentation, take the time to translate everything. Don't take for granted that everybody in the crowd will understand the foreign words, even if very similar. Even when their meaning is obvious, foreign words might distract the audience while they should be focusing elsewhere.

Also, pay attention to the fact that in the Anglo-Saxon world, decimals are expressed with a period/dot, as opposed to a comma.

Translating tables and graphs might involve going back to the original Excel/ Photoshop (or other) files and require some effort. But this process is likely to be worthwhile, as it will give you the opportunity to potentially fix your tables, graphs and figures to address all the other tips provided in this book!

You can certainly find other ways to express your pride about your country of origin!

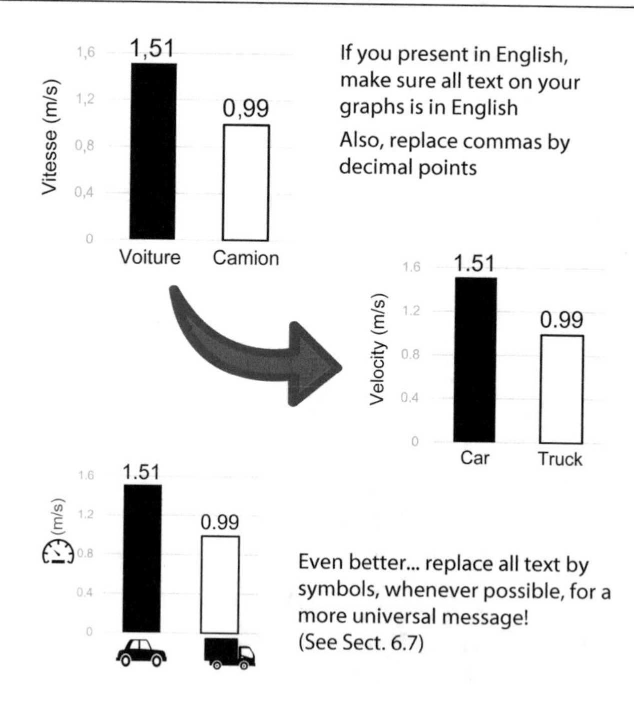

11.5 The First Slide Over and Over Again!

The approach to how you practice for your presentation is different whether you are dealing with short "time-limited" talks, or long lectures (Sect. 11.2). But in either case, the most difficult part is to get started.

It could be tempting to get going by reading something off your slide to avoid a memory blank. But as mentioned in Sect. 11.1, you should never read the content of your slide (the exceptions being citing quotes, Sect. 13.8 and acknowledgement, Sect. 9.8). Also, as mentioned in Sect. 9.1, do not waste time reading the title of your talk, or introducing your co-authors at this stage.

You must dive right into your presentation topic, concisely and without hesitation. And this requires practice. You should be repeatedly rehearsing your first slide, and possibly focusing even on the first few sentences. The first few seconds are critical. If you have to learn something by heart... it could be the first one or two sentences (but no more than that!) Focus on the first slide, rehearse it a few times until you feel very comfortable. You don't even need to have your slide in front of you.

Once you get started, the ideas will flow!

11.6 Practice Using Your Smartphone!

You should not limit yourself to practicing your talk only during dedicated sessions, comfortably seated in front of your computer. Any occasion is good to rehearse. Even silently practicing in a public area can be quite productive. But a key to an effective practice, at least prior to having memorized the content of your slides, is to have your slide deck accessible. You could carry a printout of your slides (see Sect. 11.1), or even better, have your slides loaded to your tablet or smartphone.

Some smartphones will allow you to run a presentation with all embedded animations. But if you do not have a slide presentation software installed, send a PDF version to your phone as an email attachment. A PDF version will not run animations (unless you used Option 1, see Sect. 3.2), but that should still be good enough for a good rehearsal.

Other uses for your smartphone:

- Use it as a timer, to practice for time-limited presentations
- Use it to record yourself (voice only or video) for further enhancements

If you are like most people, your smartphone is never too far! Take advantage!

11.7 Do Dry Runs with Friends and Peers

A full dry-run might not be practical for a long lecture, but for short, time-limited presentations (e.g. conferences), a dry-run in front of peers is effective, for a few reasons:

- The level of stress is more representative of the actual presentation than if you only practice on your own
- You get real-time feedback on clarity and pace of speech, quality of the slides
- It highlights ill-prepared sections of your talk (hesitations, dwelling on unimportant elements)
- You can ask for specific feedback (e.g. was this graph clear enough? Did you understand when I introduced this concept?)
- Your peers can ask you questions similar to what might be asked during the real presentation, thus enhancing your preparedness
- You can be monitored for time

Even friends or family members not familiar at all with your research topic can act as an effective dry-run audience. They should at the very least understand the objectives of your work (a good test for the quality of your introduction). And their feedback on your communication skills will be as good as anyone else's.

If a colleague presents at the same conference, practice together until you can both deliver each other's presentation!

11.8 Cut... and Cut Again!

When you are limited in time to give your presentation, you are likely to have to cut down on the number of slides and text. You will have to focus on only the most important and key concepts you want to convey. And avoid dwelling on any specific element.

Cutting slides, cutting material from your talk is usually unavoidable, and also very difficult. This is your baby; this is what you have worked so hard on. And you might feel that the audience will end up losing out if you remove some elements from your presentation. They might even realize that something is missing.

Actually… no! No one will notice that you have removed material from your original version, provided that your talk still flows. The audience will never miss slides they never knew existed. Leave some slides aside, proceed with some mourning and move on!

Aim for a shorter presentation which you can deliver slowly, rather than a long presentation you have to cram through! (see Sect. 13.6).

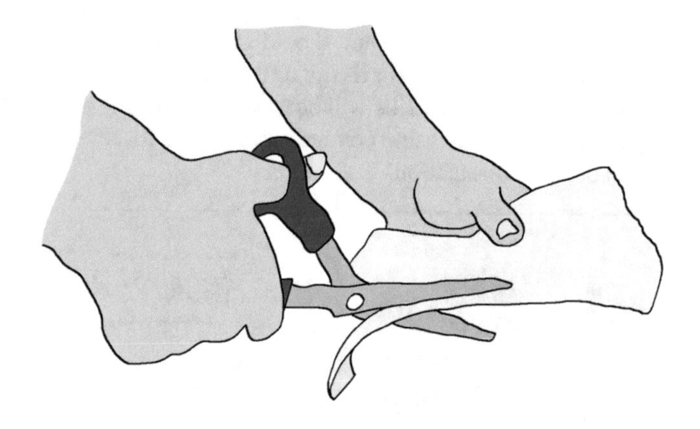

11.9 Adapt Your Slides to Your Talk

As you repeatedly practice, your talk might evolve in a direction different than what you had originally intended. As such, you might realize that the sequence of your slides no longer matches exactly how you want to say things. Do not hesitate to change the sequence, modify the animations, add some animations, remove elements, etc. Your slides have to match your talk perfectly! (not the other way around).

For instance, if you realize that you keep forgetting addressing a specific critical piece of information, make sure you add a visual cue to it in your presentation, at the exact moment you have to discuss it.

You should not need a piece of paper (cheat-sheet) to remind yourself of your talk. Your slide deck IS your cheat-sheet! If you keep forgetting about something, make sure you introduce an animated item which will remind you of discussing it at the appropriate time.

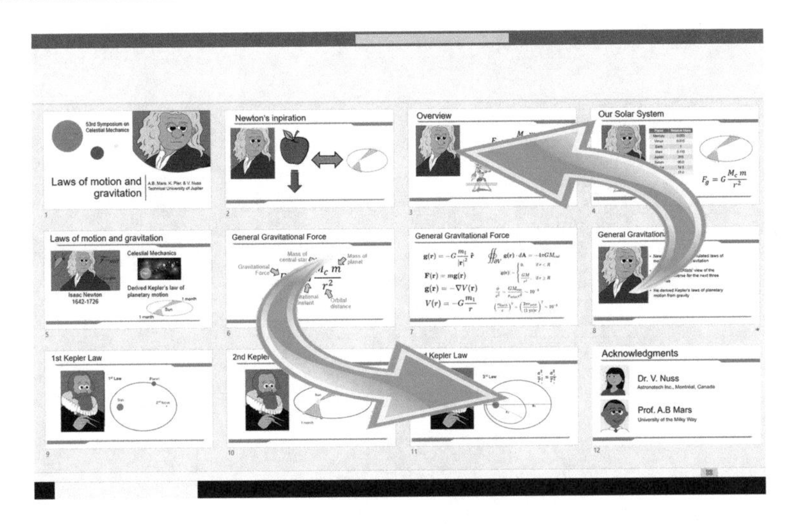

As you practice your talk, you might realize that your slides could be sequenced differently

Do not hesitate to make changes for a perfect flow

Planning 12

J.-P. Dionne, *Presentation Skills for Scientists and Engineers*,
https://doi.org/10.1007/978-3-030-66069-7_12

12.1 More Than One Presenter?

I have attended a few talks presented by two speakers, and I have never been impressed with the result. Typically, the silent speaker looks dumb doing nothing on stage (which is distracting) while his/her colleague is busy going through the slides. Moreover, the audience feels frustration when the best of the two speakers leaves the podium to the weaker one.

When considering having two presenters, ask yourself "what's in it for the audience?" Giving both speakers their 5 minutes of glory is no good enough reason.

A potentially acceptable ground for having two speakers is when both are subject matter experts in their own distinct and relevant disciplines. But even then, I still believe that having one presenter is better. The other one can simply be made available for the question period.

In a play, actors interact together, so it makes sense to have more than one actor on stage at any given time. But unless you plan your talk to include conversations among the two speakers (which would definitely be original), you should stick to only one speaker.

If you insist on having two speakers, at the very least, minimize any back and forth between the two of you.

12.2 Explore Alternative Software

All examples and references to slide presentation software from this book are based on PowerPoint. This is not surprising, given its widespread use, especially within the relatively conventional scientific community. But if conference organizers allow it, I encourage you to explore alternative slide presentation software and possibly find one that suits you better. Many of these options are "cloud- based", making it easy to work in teams and facilitating review. A quick "Google" search will give you a number of options (many are free) to choose from. Examples include Canva, Google Slides, Keynote, Prezi, and many more. See a few descriptions below.

Prezi is interesting as it allows for so-called "non-linear" presentations, where the presenter can pan and zoom from one page to the next, without it looking like a conventional "change of slide". Most of the options allow for various design and professional templates and graphics. However, while such features might enhance the appeal of your slide deck, beware of creating distractions to the audience, with the emphasis shifting from the content to the "container"!

Take a look, try out things, but keep in mind all the tips and advices from this book, which apply no matter which slide presentation software you select!

PowerPoint is the "de facto" standard for slide presentation, but it is worth exploring for alternatives

Google Slides, a web-based application, allows users to create and edit presentations online while collaborating with other users in real-time

Keynote is an Apple product which offers themes that allow users to keep consistency in colors and fonts, as well as powered 3D slide transitions for flipping pages or other transitions

With Canva, Users can choose from many professional designed templates, and edit the designs and upload their own photos through a drag and drop interface

Prezi is a visual storytelling software that features a map-like overview that lets users pan between topics, zoom in on details, and pull back to reveal context

12.3 Use Your Own Laptop?

In most instances, you are either mandated to use your own laptop or use the computer system provided by the audio-visual team onsite. But if you have the choice, what should you do?

Advantages of using your own laptop:

- Your presentation format will not be altered in any way
- The videos will run well
- You can use specialized software installed only on your laptop
- You can use your own remote tool if you have one
- You can touch up your presentation until the very last minute

On the other hand:

- You might face connection issues with the audio-visual system (especially if you use a Mac!) – time will be lost
- More time is lost switching to and from your system (under pressure from the crowd)
- You cannot use the remote tool provided by the conference
- You have to walk around with your laptop
- You must ensure to have enough battery, or a power cable

You should never underestimate the potential connection issues between your laptop and the audio-visual system. For this reason alone, I always prefer only bringing a USB stick (plus a backup, physical or "in the cloud") and use the system provided.

12.4 Dry Run: Rule Out Technical Issues

Hopefully you have the opportunity to do a dry run (without any audience) with the audio-visual specialist prior to your actual presentation.

If you are offered that opportunity, go through your presentation in "presentation mode", to make sure that:

- The screen ratio is good
- Your slides get fully displayed, without distortion
- All your movies run perfectly

You can even take advantage of that last verification to make final touch-ups and modifications to your final presentation file.

This will also allow you to familiarize yourself with the setting (podium, view of the audience, remote control if any, etc.)

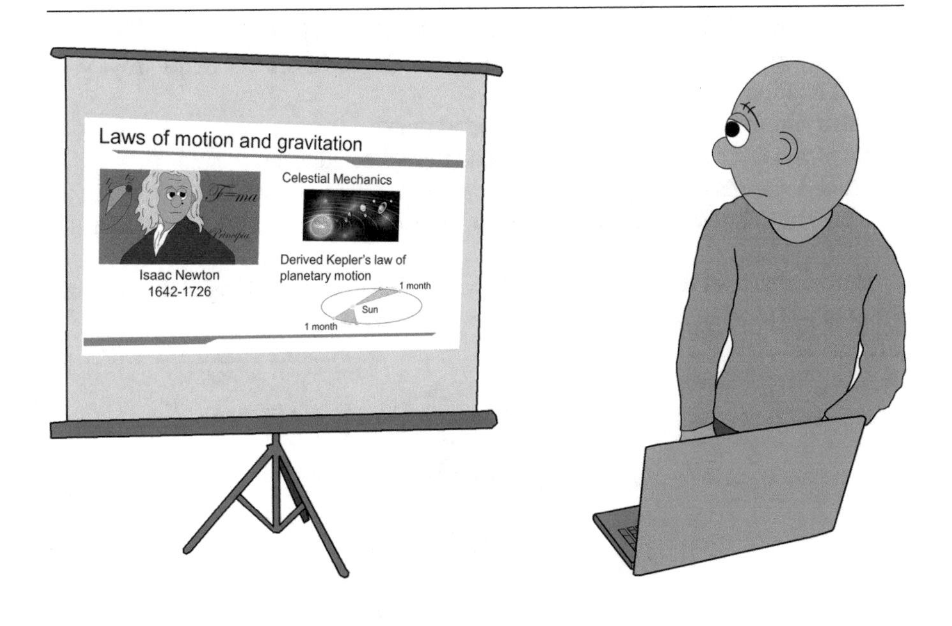

12.5 Go Through a Checklist!

To further increase your confidence level (especially if you are novice presenter), you might consider putting together a "checklist" of things to do maybe a day or two prior to your talk. At that point, it is too late to introduce major changes to your slide deck, but some fine-tuning should still be possible.

- Do you have a backup of your slide deck file handy (USB stick, cloud)?
- Have you confirmed the exact date and time of your presentation? The allotted duration (including or excluding the question period)
- Should you sit somewhere specific just prior to your talk?
- Until when can you still make changes to your file?
- Have you done a dry run (Sect. 12.4)? (focus on movies)
- Have you familiarized yourself with the room settings, microphone and remote control (see Sects. 13.1 and 13.3)
- If you use your own laptop, do you have enough battery?
- Have you sufficiently practiced the first slide? (Sect. 11.5)
- Have you clearly highlighted in your slides the most critical items from your presentation?
- Have you timed yourself to fit the allotted time?
- Have you anticipated possible questions from the audience?
- Should you make changes to your slides or to what you intend to say, based on other presentations you have seen so far, or discussions held with other scientists?

Have a checklist ready with some the items above, and anything else that is relevant to you. And then make sure all items get crossed out!

12.6 Slides for Day-to-Day Meetings

This book focuses on scientific presentations in front of crowds but slide decks can be used in many more circumstances, including informal day-to-day meetings, during which you may very well remain sitting on a chair, sharing the "attention" with a few other colleagues.

Regular meetings typically involve having to navigate through multiple file types on screen (Excel, Word, others). But rather than frantically switching back and forth between software platforms, proactively prepare for meetings with a single all-encompassing slide deck.

Put together a well-defined storyline by inserting key screen captures from the various software packages, thereby saving time switching back and forth between files. Include links to the native files, to consult them as needed for a deeper analysis.

For day-to-day meeting presentations, while I suggest you still follow most of the general advice from this book, no need to do it too strictly. Include more details (not too much) and allow yourself to add more text on the slides. The key point is to come to meetings well prepared. Putting a presentation together is a great way to achieve that goal. All other meeting attendees will benefit!

12.7 Slides for Teleconferences

At the time of writing this book (2020), the world is going through a major pandemic forcing countless employees to work from home and rely on teleconferencing. While teleconferencing platforms allow transmitting the user's video image, they also allow sharing computer screens. As such, teleconferences are not that different from regular "on-site" meetings. Therefore, all the advice provided in Sect. 12.6 (Slides for day-to-day meetings) apply just as well for conference calls involving video (e.g. feeding all relevant information in a single all-encompassing presentation file).

The fact that all the visual information goes through the screen, as compared to regular meetings where actual printed documents can be shared and physical interactions are possible, makes it even more critical to optimize what is displayed on the screen during teleconferences.

To stop your fellow co-workers getting distracted by their cat, mending their toenails or work on something completely different while you are presenting your views, it is imperative that you prepare a good visual "show", hence, a presentation. Also, given the absence of physical interaction, your presentation must be built in a way to facilitate highlighting the key points (you can't just point at the screen!)

13 Delivery

J.-P. Dionne, *Presentation Skills for Scientists and Engineers*,
https://doi.org/10.1007/978-3-030-66069-7_13

13.1 Be in Control!

Ensure you have control over the computer. You have to be able to change slides and control animations yourself. Avoid the very painful series of "next slide please"!

Use a remote whenever possible, making sure ahead of time you know how it works (find out during the dry run with the audio-visual technician). The best alternative to a remote is the keyboard. The space bar is the most convenient way to move up a slide (or animation), as it is easy to see and hit. The up and down arrows are convenient too.

Use the mouse only for activating videos or action buttons. Do not use it for moving up and down slides (right-clicking to show all options and choosing to move up one slide). That's just silly.

On the other hand, if you know for sure that you will NOT have control on the animations and slide changes, here is some advice:

- Limit the number of slides to a minimum
- Limit to a minimum the number of animations requiring manual control
- You might introduce "automatic animations", triggered based on time – but this requires you to talk according to the planned schedule

---

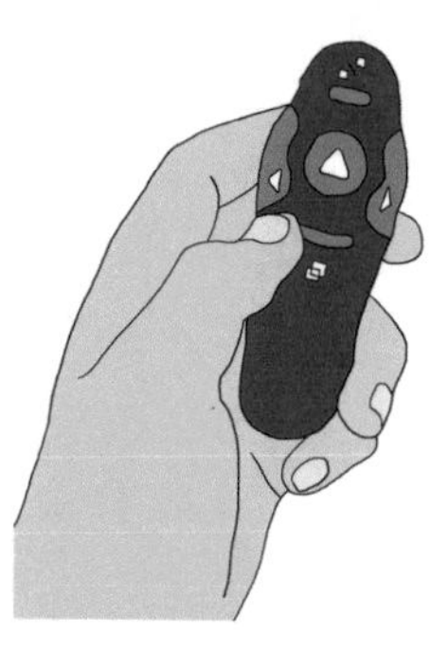

Using a remote is ideal to flip through the slides

The next best option is to use the arrow keys on the keyboard

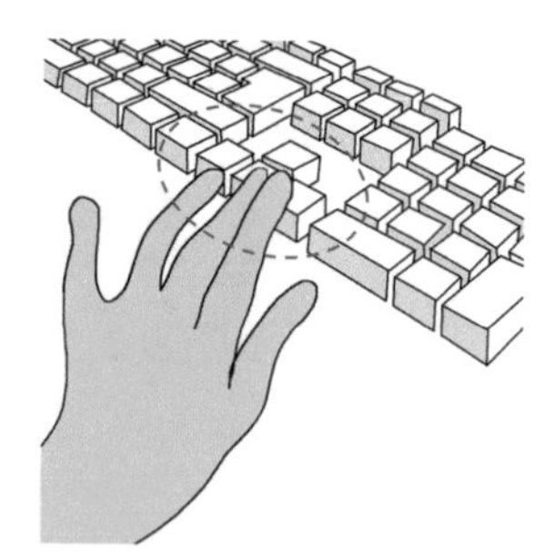

13.2 Look at the Audience

Actors in a play have a distinct advantage over actors in a movie: immediate feedback from the audience. When giving a presentation, you should also take advantage of feedback provided by the audience. Their body language will let you know whether:

- Your presentation is boring (people sleeping, looking at their phones)
- Your voice tone is too low (people sleeping)
- Your presentation is too complicated (people stop looking and listening)
- You catch their interest (everybody is staring at you/the screen)
- Your jokes (if any) are actually good (or bad) – Sect. 13.9

You can then adapt your delivery according to the feedback received (raise your voice, take time to better explain, skip to the next topic, etc.)

You also establish a better connection with the audience through a more active presentation. As such, they are more likely to pay attention to what you say and do (no one likes to be caught drowsing). Plus, you get to see who leaves the room, comes in, actively listens, etc. A wealth of information!

Looking at the audience shows that you care, rather than just going through the motion.

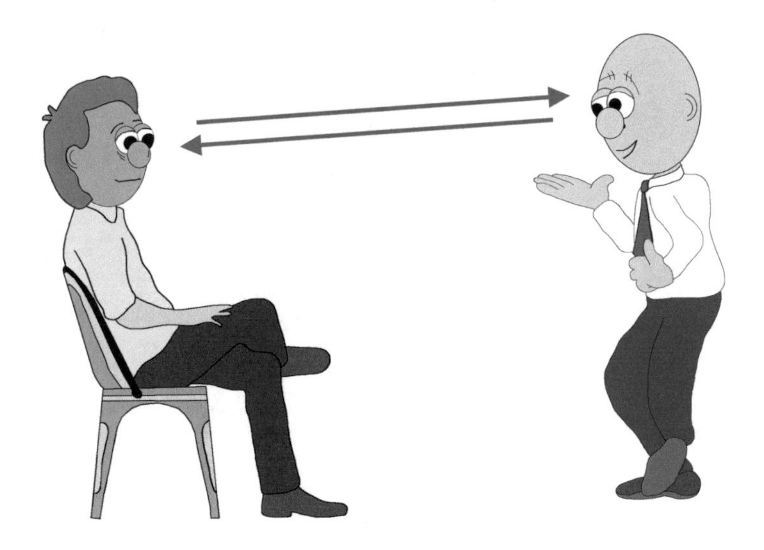

13.3 Use the Microphone

Even if you have a powerful voice, don't show off: use the microphone provided!

Irrespective of your voice strength, the volume is always louder for the front row than the back one. Loudspeakers do more than just amplifying; they distribute sounds more uniformly (if properly laid out of course). Simply adjust your voice level. Nothing wrong with talking softly!

The ideal microphones are those that get attached to your clothing, as they provide freedom to move around without affecting the projected voice. Be careful though: if you turn your head to look at the screen behind you, you will no longer be talking straight in the microphone. In such a case, it is better to rotate your entire body towards the screen.

If the microphone is not mobile and attached to the podium, resist the temptation to move around on stage. Stand still behind the podium, and make sure that the microphone position is customized for you. And don't talk with your head turned away from the microphone!

13.4 Avoid the Laser Pointer

Even a very steady person may have a hard time pointing steadily at a fixed spot. It is even worse for a nervous speaker. Bad things that happen with a laser pointer:

- The red dot goes all over the place very quickly
- Not enough time to see where the speaker wanted to aim
- The speaker leaves the pointer on (which becomes a threat)

If your presentation is well built, you should NEVER have to use a laser pointer. If you want to emphasize areas of your slides, e.g. specific spot on a photo, specific data point on a graph, etc., overlay instead geometrical figures (the red circle works in most cases). Animate the circle (no fill) to have it appear exactly when you need it, and to emphasize exactly what you need to emphasize.

Advantages:

- Always steady
- Red circle appears exactly when needed, exactly where needed
- It can stay as long as needed (can remove it with animations)
- Multiple areas can be highlighted at the same time (vary colors)
- Choice in size/shape that best fits what you want to emphasize

Another way to emphasize areas of your slides, is to use geometrical figures with a fill, and adjust the transparency.

Alternatively, you can use the PowerPoint built-in cursor (the top- left pointing arrow) to highlight specific items on your slide, but…

- The arrow cursor takes a few seconds before appearing, and quickly disappears (the cursor can be setup to be always visible, but this becomes a nuisance when you don't need it)
- The cursor-arrow comes in only one shape and size (quite small)
- You can only emphasize one item at a time

There is also the old-fashioned alternative of pointing directly at the screen, provided there is only one screen, and it is close enough to the podium!

Emphasis again: avoid the laser pointer. Anytime you feel like using one, find alternatives within your presentation itself, so that you no longer need the laser pointer.

One notable potential exception is the question period. In providing your response, you might use the laser pointer to highlight items from your slides which you had not highlighted previously.

It is quite hard to be steady when using the laser pointer, plus, very often, the audience does not have enough time to see where the speaker is pointing at

As a much better alternative to the laser pointer, introduce instead (and animate) geometrical figures to highlight what needs to be highlighted

°C	12	24	8	5
°C	29	40	24	22
mm	8.4	29.0	19.6	26.2

If documentaries do not make use of laser pointers, there is no reason for you not to be able to live without them!

13.5 Euh...

Ideally, you must avoid as much as possible the filler type words, especially the "euh" which indicates you have momentarily forgotten what you wanted to talk about.

Or, you might have the wrong habit of starting every sentence and every slide with the same word or words (e.g. "so..."). Trick: make a conscious effort, when you flash a new slide, not to say anything, and stay silent for a second or two. Then, make sure you start your sentence using only meaningful words. Dive right into what you need to say.

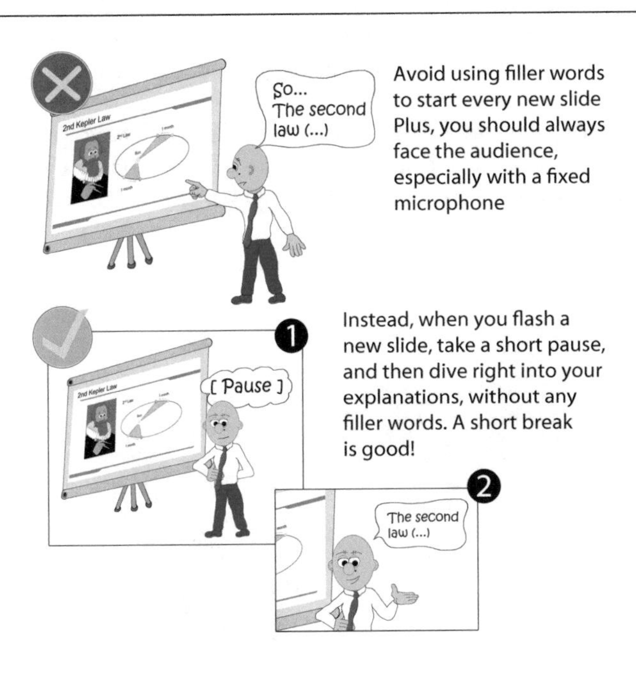

13.6 Take Your Time

Always leave enough time for digesting the entire content of each slide (which should be easy if you have not included text and have carefully introduced graphs). Do not hesitate to stop for a few seconds once in a while, for instance, when transitioning to a new slide.

But of course, to have the luxury of taking your time, you must have carefully planned the amount of material to be conveyed if the allotted time is very tight.

Resist to the temptation of including every single aspect of your work in your talk. You will be the only one to know if you chose not to present something (see Sect. 11.8). No one will miss what you have not included. Just provide the right amount of details, be concise, do not dwell on anything without a good reason.

You will also appreciate the luxury of taking your time and not feeling pressed by the clock.

When preparing and practicing for your oral presentation, you have to keep the duration of your talk in mind. But once on stage, talk duration should be the least of your concerns, since you will have made all required efforts ahead of time to ensure timing is good.

13.7 Insert a Blank Slide

Not everything you talk about necessarily requires visual support. Whether it is at the very beginning of your talk, or somewhere right in the middle, there is nothing wrong about inserting a blank slide. That is likely to shock the audience a bit, as they will immediately think about an audio-visual problem. But as you keep talking despite of the blank page, they will now turn your attention to YOU. And ideally, you should be looking at the audience at the same time. This blank slide is thus an ideal time to connect with your audience. Don't forget that any presentation is an opportunity to market yourself. Having the audience look directly at you rather than always staring at your slides is a great idea.

A blank slide should ideally have the same background as the rest of your slides, that is, if you have a generally white background, that blank page should be white. If you work with a blue background, then your blank page should be blue. Otherwise, it might create too much of a contrast and a break from your talk.

Once in a while, documentaries will simply show a zoom on the show host introducing a topic. This is the equivalent of a blank slide.

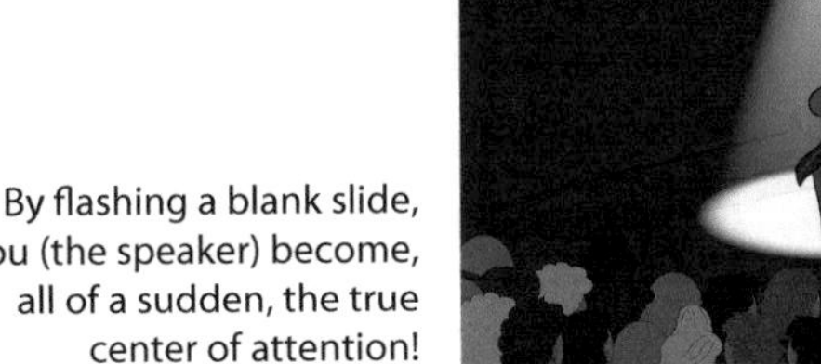

By flashing a blank slide, you (the speaker) become, all of a sudden, the true center of attention!

13.8 The Art of Citing Quotes

Citing a quote is the only case where it is OK to have text written on the screen and reading it out loud.

- Only show the quote as you start citing it. Not before (otherwise the audience will start reading it ahead of you), and not after (what would be the point?)
- You can read a quote at a relatively slow pace, as it appears on the screen
- For a long quote, let it appear in sections, so that readers do not get ahead of you, and the amount of text is not overwhelming

---

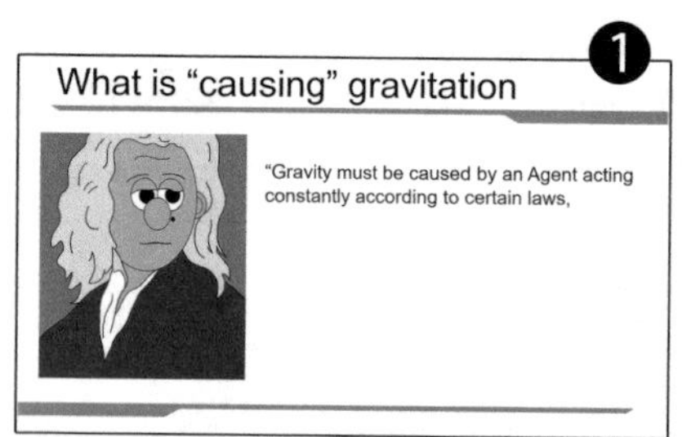

Do not show entire citations all at once, especially if they are rather long

Animate them as you speak

13.9 Joking Around... Really?

Any topic, even serious ones, can benefit from some humor. Humor will entertain the audience, raise their attention level and might even help getting some points across. Humor can make your presentation livelier.

Given all this, you should include humor in your presentation, right?

Actually, my STRONG advice is to avoid humor, unless joking around is one of your personality traits, or if you have used humor successfully and repeatedly in the past. If you are not a "natural", just leave the jokes aside.

But if despite not having a proven "entertainer" record, you still insist on including humor, here is my advice:

- Do not expect a reaction from the audience – be ready for, and even plan for a total indifference of the audience to your joke
- Do not provide non-verbal cues that you are joking. Do it covertly
- Do not laugh at your own joke, especially if you are the only one laughing
- Do not let your "joke" alter the flow of your presentation – whether the audience reacts or not, keep going!

We can all relate to some presenter being embarrassed after telling a poor joke. Avoid the embarrassment... An interesting talk does NOT need to be funny.

13.10 Follow the "TED Recipe"?

"TED talks" are popular presentations available online (www.ted.com) addressing multiple topics within the areas of science and culture. Their distinctive format is characterized by a time limit (18 minutes) and only light reliance on visuals. Their focus truly is on the speakers, who get specific training to make their talks as innovative and engaging as possible.

Should you follow this TED format? You should definitely aim at connecting with the crowd, as the TED speakers do. Along these lines, I had suggested in Sect. 13.7 to insert a blank slide, whenever visuals are not needed, precisely for the purpose of better connecting with the audience. TED talks push that concept to the extreme.

But when presenting scientific findings to experts in your field, graphs, tables and other illustrations are critical to go beyond just scratching the surface. And in general, it is desirable to excite as many senses as possible during a presentation. As such, adding visuals is key, and the current book provides you a wealth of tips and advice in that respect.

Nevertheless, I encourage you to watch a few TED talks to inspire yourself in enhancing your connection with the audience.

14 After the Talk

J.-P. Dionne, *Presentation Skills for Scientists and Engineers*,
https://doi.org/10.1007/978-3-030-66069-7_14

14.1 Repeat/Rephrase All Questions

If a microphone is not provided for the audience, or if someone is too lazy to walk up to the microphone or too anxious to wait for a microphone to be provided, there is a risk that a question asked to you will not be heard properly by the entire crowd.

In addition, it could be that a question is not properly worded or a bit confusing (the audience does not prepare as well as the speakers do). Or it could be that a single person will ask a few questions at once, and you have to focus on one at a time.

As such, I suggest that you consider repeating or rephrasing the question in your own microphone, so that it will be heard by all. In addition, rephrasing the question allows you to check with the person who asked it whether you have understood it properly (body language typically provides sufficient confirmation).

Rephrasing might also allow you to give a slightly different spin to the question, moving away from something you are not too comfortable with, towards an area of greater confidence!

Repeating or rephrasing questions demonstrate a level of respect to the audience.

14.2 Leaving Your Slide Deck Behind

You might be asked to leave a copy of your slide deck behind. However, your slide deck is likely to have been optimized for an oral presentation, not towards a standalone document to be consulted in your absence. To address this, create a distinct version of your presentation in PDF format, specifically aimed for leaving behind as a document:

- PDF files do not support animations (except Type 1 animations, Sect. 3.2). Make sure that your animated items do not cause interferences or other issues in your PDF file
- If you have Type 1 animations (animations appearing on multiple slides), consider only showing the final result
- Movies will appear as images in your PDF (first frame from the video). Consider replacing this image by the most relevant frame in your video, and make it clear this originally was a video
- If a video was critical, turn it into a series of still images (Sect. 5.3)
- Remove any slides too sensitive to leave behind
- Consider adding slides to include information only provided verbally during your talk
- Consider adding text, since this document is no longer meant to be provided as an oral presentation, but rather aimed at being read

---

14.3 Also Good for Reports

This book has emphasized numerous times the need to adapt graphs and tables from reports (scientific, financial or others) for the purpose of a presentation. What is acceptable for a report, is not necessarily acceptable for an on-screen live presentation. Specifically, data points and fonts must be large enough, graphs must be clean and include only the required information, and colors must help to distinguish the different data sets.

But looking at the corollary, what is good for a presentation will typically be quite good as well for a written test or financial report: readers of a report will also benefit from larger data points on graphs, cleaner set of data, legends directly applied to data sets, etc. As such, whatever you have learned in this book regarding graphs and tables for presentations should also be applied to graphs and tables to be included in written reports.

One distinction though: written reports do not benefit from a presenter introducing and explaining graphs and tables in real- time. As such, reports must be more 'self-contained'. For instance, using icons in graphs and tables instead of text, while still possibly OK, might not be as appropriate for test reports. Maybe a combination of both text and icons could be useful. Use your best judgment!

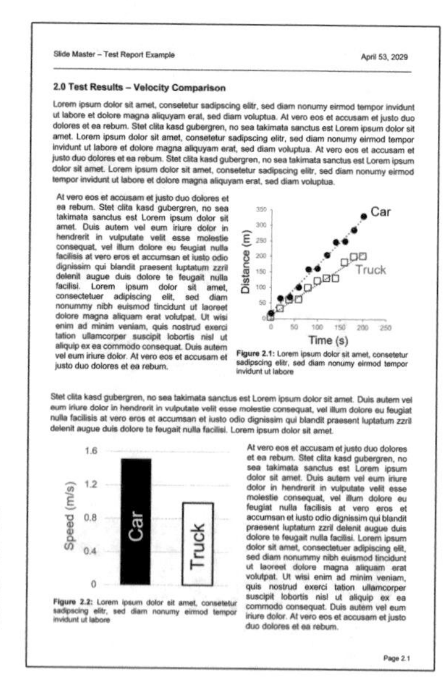
Slide Master – Test Report Example   April 53, 2029

2.0 Test Results – Velocity Comparison

Lorem ipsum dolor sit amet, consetetur sadipscing elitr, sed diam nonumy eirmod tempor invidunt ut labore et dolore magna aliquyam erat, sed diam voluptua. At vero eos et accusam et justo duo dolores et ea rebum. Stet clita kasd gubergren, no sea takimata sanctus est Lorem ipsum dolor sit amet. Lorem ipsum dolor sit amet, consetetur sadipscing elitr, sed diam nonumy eirmod tempor invidunt ut labore et dolore magna aliquyam erat, sed diam voluptua. At vero eos et accusam et justo duo dolores et ea rebum. Stet clita kasd gubergren, no sea takimata sanctus est Lorem ipsum dolor sit amet. Lorem ipsum dolor sit amet, consetetur sadipscing elitr, sed diam nonumy eirmod tempor invidunt ut labore et dolore magna aliquyam erat, sed diam voluptua.

At vero eos et accusam et justo duo dolores et ea rebum. Stet clita kasd gubergren, no sea takimata sanctus est Lorem ipsum dolor sit amet. Duis autem vel eum iriure dolor in hendrerit in vulputate velit esse molestie consequat, vel illum dolore eu feugiat nulla facilisis at vero eros et accumsan et iusto odio dignissim qui blandit praesent luptatum zzril delenit augue duis dolore te feugait nulla facilisi. Lorem ipsum dolor sit amet, consectetuer adipiscing elit, sed diam nonummy nibh euismod tincidunt ut laoreet dolore magna aliquam erat volutpat. Ut wisi enim ad minim veniam, quis nostrud exerci tation ullamcorper suscipit lobortis nisl ut aliquip ex ea commodo consequat. Duis autem vel eum iriure dolor. At vero eos et accusam et justo duo dolores et ea rebum.

Figure 2.1: Lorem ipsum dolor sit amet, consetetur sadipscing elitr, sed diam nonumy eirmod tempor invidunt ut labore

Stet clita kasd gubergren, no sea takimata sanctus est Lorem ipsum dolor sit amet. Duis autem vel eum iriure dolor in hendrerit in vulputate velit esse molestie consequat, vel illum dolore eu feugiat nulla facilisis at vero eros et accumsan et iusto odio dignissim qui blandit praesent luptatum zzril delenit augue duis dolore te feugait nulla facilisi. Lorem ipsum dolor sit amet.

Figure 2.2: Lorem ipsum dolor sit amet, consetetur sadipscing elitr, sed diam nonumy eirmod tempor invidunt ut labore

At vero eos et accusam et justo duo dolores et ea rebum. Stet clita kasd gubergren, no sea takimata sanctus est Lorem ipsum dolor sit amet. Duis autem vel eum iriure dolor in hendrerit in vulputate velit esse molestie consequat, vel illum dolore eu feugiat nulla facilisis at vero eros et accumsan et iusto odio dignissim qui blandit praesent luptatum zzril delenit augue duis dolore te feugait nulla facilisi. Lorem ipsum dolor sit amet, consectetuer adipiscing elit, sed diam nonummy nibh euismod tincidunt ut laoreet dolore magna aliquam erat volutpat. Ut wisi enim ad minim veniam, quis nostrud exerci tation ullamcorper suscipit lobortis nisl ut aliquip ex ea commodo consequat. Duis autem vel eum iriure dolor. At vero eos et accusam et justo duo dolores et ea rebum.

Page 2.1

Consider tips provided in this book for graphs and tables to include in reports as well!

15 Conclusion

J.-P. Dionne, *Presentation Skills for Scientists and Engineers*,
https://doi.org/10.1007/978-3-030-66069-7_15

15.1 Conclusion

This book includes close to a hundred tips to improve your presentations. But not all of these tips were created equal. As a conclusion, I suggest you clearly emphasize the following ones:

- Get inspired by documentaries – what they do, what they don't
- Don't compete with yourself – avoid text
- Provide visual info exactly when needed
- Animate your text, tables and graphs
- But avoid fancy useless animations
- Avoid the laser pointer – embed emphasis in your presentation
- Invest time on your graphs – don't just copy from reports
- Avoid math as much as possible
- Dive into your presentation (no outline, acknowledgments at end)
- Practice a lot but don't learn by heart
- Remove material to avoid having to rush through your slides
- No one will notice you removed something

The rest is mostly details and the 'mechanics' of it all. A few more important things to remember:

- Investing time in your slide deck demonstrates respect for your audience. People feel smarter when they understand

While you may or may not be a good speaker, a good slide deck will always make your message go through with great success!

Printed in the United States
by Baker & Taylor Publisher Services